PRACTICAL
BIOTECHNOLOGY
AND
PLANT TISSUE CULTURE

PRACTICAL
BIOTECHNOLOGY
AND
PLANT TISSUE CULTURE

PRACTICAL
BIOTECHNOLOGY
AND
PLANT TISSUE CULTURE

SANTOSH NAGAR

M.Sc. (Gold Medalist), M.Phil.
Head, Department of Botany
P.M.B. Gujarati Science College,
Indore (M.P.)

Dr. MADHAVI ADHAV

M.Sc. (Gold Medalist), Ph.D., Botany
Asstt. Prof., Department of Botany
P.M.B. Gujarati Science College,
Indore (M.P.)

S. CHAND & COMPANY LTD.
(AN ISO 9001 : 2000 COMPANY)
RAM NAGAR, NEW DELHI-110 055

S. CHAND & COMPANY LTD.
(An ISO 9001 : 2000 Company)
Head Office: 7361, RAM NAGAR, NEW DELHI - 110 055
Phone: 23672080-81-82, 9899107446, 9911310888
Fax: 91-11-23677446
Shop at: **schandgroup.com**; e-mail: **info@schandgroup.com**

Branches :

AHMEDABAD : 1st Floor, Heritage, Near Gujarat Vidhyapeeth, Ashram Road, **Ahmedabad** - 380 014, Ph: 27541965, 27542369, ahmedabad@schandgroup.com

BANGALORE : No. 6, Ahuja Chambers, 1st Cross, Kumara Krupa Road, **Bangalore** - 560 001, Ph: 22268048, 22354008, bangalore@schandgroup.com

BHOPAL : 238-A, M.P. Nagar, Zone 1, **Bhopal** - 462 011, Ph: 4274723. bhopal@schandgroup.com

CHANDIGARH : S.C.O. 2419-20, First Floor, Sector - 22-C (Near Aroma Hotel), **Chandigarh** -160 022, Ph: 2725443, 2725446, chandigarh@schandgroup.com

CHENNAI : 152, Anna Salai, **Chennai** - 600 002, Ph: 28460026, chennai@schandgroup.com

COIMBATORE : Plot No. 5, Rajalakshmi Nagar, Peelamedu, **Coimbatore** -641 004, (M) 09444228242, coimbatore@schandgroup.com

CUTTACK : 1st Floor, Bhartia Tower, Badambadi, **Cuttack** - 753 009, Ph: 2332580; 2332581, cuttack@schandgroup.com

DEHRADUN : 1st Floor, 20, New Road, Near Dwarka Store, **Dehradun** - 248 001, Ph: 2740889, 2740861, dehradun@schandgroup.com

GUWAHATI : Pan Bazar, **Guwahati** - 781 001, Ph: 2738811, guwahati@schandgroup.com

HYDERABAD : Sultan Bazar, **Hyderabad** - 500 195, Ph: 24651135, 24744815, hyderabad@schandgroup.com

JAIPUR : A-14, Janta Store Shopping Complex, University Marg, Bapu Nagar, **Jaipur** - 302 015, Ph: 2719126, jaipur@schandgroup.com

JALANDHAR : Mai Hiran Gate, **Jalandhar** - 144 008, Ph: 2401630, 5000630, jalandhar@schandgroup.com

JAMMU : 67/B, B-Block, Gandhi Nagar, **Jammu** - 180 004, (M) 09878651464

KOCHI : Kachapilly Square, Mullassery Canal Road, Ernakulam, **Kochi** - 682 011, Ph: 2378207, cochin@schandgroup.com

KOLKATA : 285/J, Bipin Bihari Ganguli Street, **Kolkata** - 700 012, Ph: 22367459, 22373914, kolkata@schandgroup.com

LUCKNOW : Mahabeer Market, 25 Gwynne Road, Aminabad, **Lucknow** - 226 018, Ph: 2626801, 2284815, lucknow@schandgroup.com

MUMBAI : Blackie House, 103/5, Walchand Hirachand Marg, Opp. G.P.O., **Mumbai** - 400 001, Ph: 22690881, 22610885, mumbai@schandgroup.com

NAGPUR : Karnal Bag, Model Mill Chowk, Umrer Road, **Nagpur** - 440 032, Ph: 2723901, 2777666 nagpur@schandgroup.com

PATNA : 104, Citicentre Ashok, Govind Mitra Road, **Patna** - 800 004, Ph: 2300489, 2302100, patna@schandgroup.com

PUNE : 291/1, Ganesh Gayatri Complex, 1st Floor, Somwarpeth, Near Jain Mandir, **Pune** - 411 011, Ph: 64017298, pune@schandgroup.com

RAIPUR : Kailash Residency, Plot No. 4B, Bottle House Road, Shankar Nagar, **Raipur** - 492 007, Ph: 09981200834, raipur@schandgroup.com

RANCHI : Flat No. 104, Sri Draupadi Smriti Apartments, East of Jaipal Singh Stadium, Neel Ratan Street, Upper Bazar, **Ranchi** - 834 001, Ph: 2208761, ranchi@schandgroup.com

VISAKHAPATNAM : Plot No. 7, 1st Floor, Allipuram Extension, Opp. Radhakrishna Towers, Seethammadhara North Extn., **Visakhapatnam** - 530 013, (M) 09347580841, visakhapatnam@schandgroup.com

First Edition 2009

ISBN: 81-219-3200-9   **Code** : 03 381

PRINTED IN INDIA
*By Rajendra Ravindra Printers Pvt. Ltd., 7361, Ram Nagar, New Delhi -110 055
and published by S. Chand & Company Ltd., 7361, Ram Nagar, New Delhi -110 055.*

Dedicated to
Shri Saibaba
and
Our Parents

Dedicated to

Shri Satbaba

and

Our Parents

ACKNOWLEDGEMENT

Writing is a lonely activity. We never truly alone however our family was always there for encouraging and cheering. We wish to thank them all. Special thanks to readers of the book.

We express our deep sense of gratitude to our teacher for having inspired me to undertake this writing project. We are gratefull to our colleagues and friends for making valuable suggestions.

We must also record our indebtness to friends who provided their strong moral support throughout this venture.

Finally we thank to S.Chand & Company Ltd. for bringing out this book in a beautiful form in a very short duration.

PREFACE

We have a great pleasure to present the **Practical Biotechnology and Plant Tissue Culture.** Recombinant DNA technology commonly referred to as genetic engineering is one of the principal thrusts of the emerging high technologies in biological sciences. We are now experiencing a rapid shift of national priorities in research and development. As we approach the twenty first century, we see biology emerging as one of the top priorities in the field of science and among biological sciences, biotechnology has gained new slature. It is the science in preparation of new millennium.

Recently, biotechnology has been introduced in the syllabus of most of the Indian Universities at graduate and/or post graduate levels. Authors realise the problems of teachers and students during practicals. Therefore, this book is compiled to provide an excellent introduction to practicals of Biotechnology and Plant Tissue Culture for the teachers and students of Indian Universities.

The book starts with an introduction to basic knowledge of instruments which deals with principle, working, uses, limitations, and precautions of about ten instruments. This detailed text will provide a knowledge of handling instruments. For successfully performing practicals, it is very essential to have basic knowledge of preparation of culture media for bacterial growth and for plant tissue culture and also preparation of standard solution for various experiments. All this has been given in the book in simple language.

The biotechnological exercises such as Plasmid and DNA isolation, DNA size determination, Bacterial transformation, Restriction digestion, Purification of DNA, Bacteriophage titration, PCR, Gus gene assay, RFLP, RAPD, Isolation of *Agrobacterium* spp. along with some basic techniques in microbiology and biotechnology like Isolation of bacteria by Streak and Pour plate method, Growth characteristics of *E.coli.* by Plating and Turbidimetric method, have been written by the authors in simple and lucid style. In plant tissue culture exercises, the emphasis is given on Protoplast isolation, Viability testing, Protoplast fusion, Callus culture, Cell suspension culture, Androgenesis, Somatic embryogenesis, Preparation of artificial seeds, Organogenesis, Isolation of plant DNA and Transfer of plantlet to greenhouse field.

The authors practically performed all the experiments, and have then written this book. All the figures have been drawn by the authors themselves and flowcharts of experiment is also drawn for easy understanding. Special care has been taken to present the subject matter in a student friendly manner.

In addition to the above mentioned exercises, we have also given viva-voce material.

We hope that the book will be useful to teachers as well as students.

The authors shall welcome suggestions from the readers.

Santosh Nagar
Dr. Madhavi Adhav

CONTENT

Experiment **1**

OBJECT : Principles, Working and uses of Instruments.

1. AUTOCLAVE

An apparatus using pressurized steam for sterilization. The killing action of heat with time temperature relationship is useful in controlling microbial population to the desired level. Practical procedures by which heat is employed are conveniently divided into two categories moist **heat** and **dry heat.** In autoclave moist heat is used for sterilization.

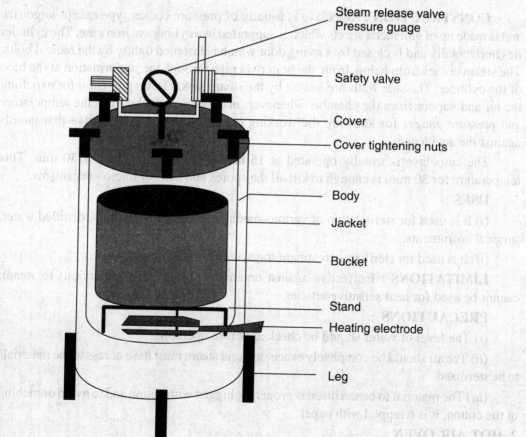

Steam release valve
Pressure guage
Safety valve
Cover
Cover tightening nuts
Body
Jacket
Bucket
Stand
Heating electrode
Leg

Fig. 1. Autoclave.

PRINCIPLE : Heat in the form of saturated steam under pressure is used for sterilization. Steam under pressure provides temperature higher than boiling as shown in Table. It has the advantages of rapid heating, penetration and moisture in abundance, which facilitates the coagulation of proteins.

Relationship betwzeen pressure and temperature.

Steam Pressure1b/inch^2	Temperature°C
0	100.0
5	109.0
10	115.0
15	121.5
20	126.5
25	130.5
30	135.5
40	141.5

CONSTRUCTION : Autoclave is usually of pressure cooker type except large size and is made up of gun metal sheets which is supported in a cylindrical iron case. The cylinder lies horizontally and is closed by a swing door which is fastened tightly by the radical bolts. The steam comes from below from the boiler via pipe through the performation at the base of the cylinder. The side walls are heated by the steam jacket it has provision for expelling the air and vapour from the chamber whenever necessary. Autoclave has the temperature and pressure gauges for knowing the working condition. It has safety valve that guards against the accidents.

The autoclave is usually operated at 15 lb/inch^2 steam pressure for 30 min. This temperature for 30 min. is enough to kill all the spores and cells of micro - organisms.

USES :

(*i*) It is used for sterilization of various media, normal saline, buffers, distilled water, surgical instruments.

(*ii*) It is used for sterilizing treatment trays, utensils and other liquids.

LIMITATIONS : Ineffective against organisms in materials impervious to steam; cannot be used for heat sensitive articles.

PRECAUTIONS :

(*i*) The level of water should be checked before operating.

(*ii*) The air should be completely evacuated and steam must have access to the materials to be sterilized.

(*iii*) The material to be sterilized is properly plugged with cotton and to avoid drenching of the cotton, it is wrapped with paper.

2. HOT AIR OVEN

Dry heat sterilization is recommended where it is either undesirable or unlikely that steam under pressure will make direct and complete contact with the material to be sterilized. The apparatus employed for this type of sterilization may be special electric or gas oven.

PRINCIPLE : Here, sterilization is done by the application of dry heat. It kills the organisms by oxidizing their chemical constituents. It is less effective than moist heat and also less penetrating.

Hot air oven is double walled chamber, with the gap between two walls is properly insulated. It is electrically heated from below and the heating elements are so arranged that the whole of the chamber is uniformly heated. There is a in-built thermostat when required, it helps in regulating the temperature. The calibration knob sets the desired temperature. For sterilization the holding time depends upon the temperature. If the temperature of oven is 160°C, the holding time should be 1 hour. At 180°C it should be 30 minutes and at 140°C it is 90 minutes.

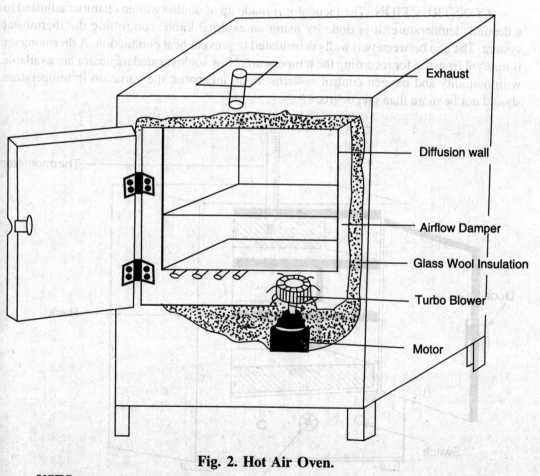

Fig. 2. Hot Air Oven.

USES :

(*i*) It is used for sterilizing materials impermeable to or damaged by moisture *e.g.*, oils, glass, sharp instruments, metals.

(*ii*) It is also used to maintain sterility of already sterilized glassware.

LIMITATIONS : Destructive to materials which cannot withstand high temperature for long periods.

PRECAUTIONS :

(*i*) The glass materials should be wrapped and dried before keeping inside the oven, otherwise it may break.

(*ii*) After holding time is over, time must be allowed before the oven is opened or else sudden rush of cool air inside may result in explosion and breaking of glassware.

3. INCUBATOR

The temperature greatly influences the microbial growth. Therefore, instrument is generally designed called incubator that can allow the desired micro-organisms to grow at a particular temperature. It is operated to allow the microbial growth on a suitable medium under appropriate temperature.

CONSTRUCTION: The incubator is made up of double walled chamber adjusted to a desired temperature. It is done by using an external knob controlling the thermostat system. The gap between two walls is insulated to prevent heat conduction. A thermometer is inserted from top for recording the temperature. Now sophisticated incubator are available with humidity and oxygen control systems. In an incubator, the variation in temperature should not be more than one degree.

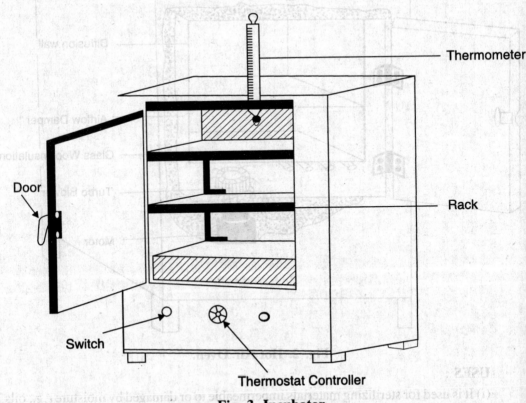

Fig. 3. Incubator.

USES : For growth of desired micro-organisms at particular temperature.

PRECAUTIONS:

(*i*) The door of the incubator should be opened only when necessary.

(*ii*) It should be cleaned and sprayed with disinfectant regularly otherwise due to constant use the environment of the chamber becomes contaminated with different micro-organisms.

(*iii*) The petridishes are incubated in the inverted position, otherwise the water of condensation may fall and disturb the colony formation.

(*iv*) If the tubes are to be incubated for a long time or at high temperature, the medium may become too dry due to excessive evaporation. In such cases cotton plug should be pushed inside the neck of the tube and the tube should be covered by rubber cap so as to cover the plug.

4. LAMINAR AIR FLOW

Biological safety cabinets employing High Efficiency Particulate Air (HEPA) filters, remove 99.97% of 0.3 μm particles, are one of the most important air filtration system.

It involves sucking in off-air of the room or cabinet and blowing out the air through a bank of filters with uniform velocity and in parallel flow line. The particles in the air are trapped on the filter medium by electrostatic attraction as well as by mechanical trapping. Inside the chamber one fluorescent tube and other UV tubes are fitted. Two switch for these tubes and a separate switch for regulation of air. The air flow are fitted outside the apparatus. Both horizontal and vertical laminar air flow systems are now used in microbiological, biotechnological and pharmaceutical laboratories.

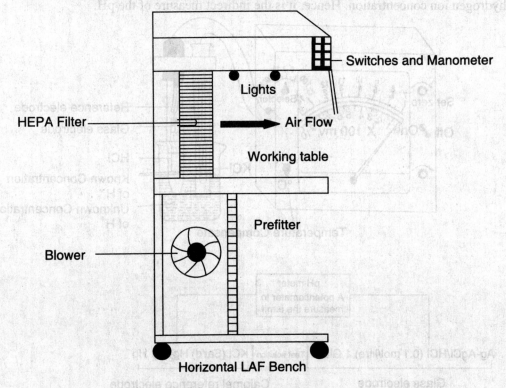

Fig. 4. Laminar air flow.

USES :

(*i*) It is used for inoculation, transfer of culture and in opening the lyophilized culture.

(*ii*) For sterilizing heat labile media components, antibiotics and when the sterile working surface is needed for conducting a wide variety of assays, preparation of media, examining or culturing the microbes and tissue cultures.

PRECAUTIONS :

(*i*) Initially, dust particles are removed from the surface of the laminar air flow with the help of smooth cloth containing alcohol.

(*ii*) Switch on the UV light for a period of 30 minutes before starting experiment so as to kill the micro-organisms if any present in the area of working space.

(*iii*) One should not talk inside the chamber while performing microbial culture transfer.

3. pH METER

It is used for the measurement of pH (acidity or alkanity of a solution) of solution of unknown pH as well as for setting of pH of various media used for cultivation and testing biochemical activities of micro-organisms.

COMPONENTS : Pair of Electrode and electric circuit. It consists of an electrode pair and an electric circuit. The electrodes are very sensitive to any change of hydrogen ion concentration of any medium or fluid. This pair is a glass electrode (Indicator electrode) and calomel electrode (Reference electrode). The electric circuit measure the electromotive force (e.m.f.) developed across the electrode pair and this e.m.f. develops depends upon the hydrogen ion concentration. Hence, it is the indirect measure of the pH.

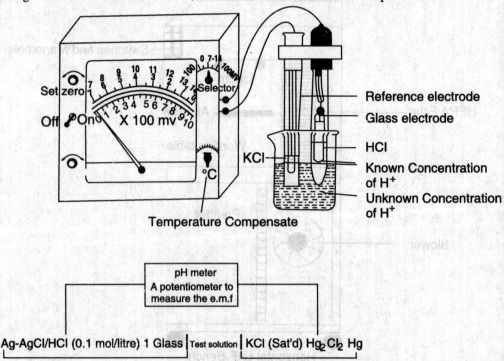

Fig. 5. pH Meter.

The glass electrode consists of a very thin bulb at one end. The bulb is made up off high onductivity glass and is sensitive to pH change. The bulb contains HCl solution (0.1 N) nside it and is connected to a platinum wire via a silver - silver chloride electrode which is eversible to hydrogen ion concentration.

The calomel electrode contains saturated solution of KCl and crystals of KCl inside it nd is connected to platinum wire via calomel paste and mercury. The pH meter is calibrated efore use by means of standard solution whose pH varies at different temperature.

USES :

(*i*) It is used to determine pH of solution of unknown pH.

(*ii*) For setting the pH of various media used for cultivation and testing biochemical ctivities of micro-organisms.

PRECAUTIONS :

(*i*) Extreme care should be taken to handle the electrodes.

(*ii*) Never allow the bulb of the glass electrode to touch the beaker.

(*iii*) Before measuring any pH, ensure that calomel electrode is filled with saturated olution of KCl.

(*iv*) Standardize the pH meter frequently against standard buffer solution of known pH alue which must be as near as the pH to be measured.

(*v*) Always wash the electrodes with distilled water between different measurements.

(*vi*) It is necessary to give the electrode time to reach equilibrium to avoid drift in the eading.

(*vii*) Care should be taken not to remove the electrode while the measuring circuit is losed.

(*viii*) When the instrument is not in use keep the glass electrode immersed in water.

. SPECTROPHOTOMETER

The spectrophotometer is used to measure absorbance experimentally. This produces ght of a preselected wavelength, directs it through the sample (usually dissolved in a solvent nd placed in a cuvette) and measures the intensity of light transmitted by the sample.

PRINCIPLE : It is based on Lambert and Beer's law. When light (monochromatic or eterogenous) is incident upon a homogenous medium a part of the incident light is reflected, part is absorbed by the medium and the remainder is allowed to transmit as such. Amount f light absorbed by the solution of unknown concentration is compared to that of standard olution and the amount of unknown sample is found out.

COMPONENT : The major components are :

(*i*) **Light Sources :** For the purposes of molecular absorption measurement *e.g.*,)euterium, hydrogen lamps a continum spectrum in UV region tungsten filament lamp (Visible nd near infrared radiation).

(*ii*) **Monochromator :** (Including, various filters, slits and mirrors). It is the optical ystem of spectrophotometer to select monochromatic light.

(*iii*) **Sample Container or Sample Chamber :** The sample is placed in a tube or uvette made of glass, quartz or other transparent material (fused silica).

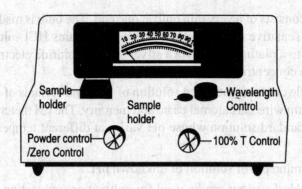

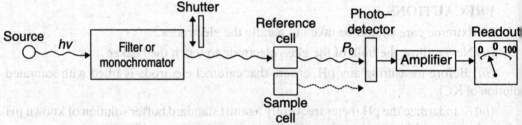

Fig. 6. Spectrophotometer.

(*vi*) **Detector** : In order to detect radiations transmitted through the sample. Three types of photosensitive devices are photovoltic cell, phototubes and photomultiplier tubes.

(*v*) **Recorder** : Meter to read the transmission by simply pushing a button, one can obtain the UV - VIS spectrum of a sample displayed on a computer screen in less than 1 Sec.

USES :

(*i*) For quantitative determination of a large variety of inorganic and organic species.

(*ii*) It is useful tool for structural elucidation.

7. COLORIMETER

Colorimeter is an instrument used for determination of concentration of compounds in solution. The concentration of the compounds present in solution is measured by the amount of light transmitted through the solution and here colour intensity is measured in terms of light absorbed.

The transmittance of light by a solution is governed by the Beer Lambert Law. In biochemical studies, determination of the concentration and identification of a substance in solution is measured by the amount of light absorbing molecules of substance present in the solution. The transmittance or absorbancy is therefore a measurement of the ratio of initial intensity (Io) of the incident rays to the intensity of light transmitted (I) out of the solution and can be expressed as Log ioto/I. The absorbency further depends upon factors such as the path length and the colour of the solution. The cell mass in suspension causes turbidity and therefore the amount of light absorbed or scattered is proportional to the mass of cells in the light path. In such case the ratio log 10Io/I can also be used to measure the optical density of the culture or the suspension.

COMPONENTS :

(*i*) A source of light.

(*ii*) Filters for selecting sufficiently narrow wave band.

(*iii*) Cuvettes with two parallel flat sides.

(*iv*) Photosensitive element - Photosensitive disc is attached with a galvanometer.

(*v*) Galvanometer which is calibrated both in terms of percentage transmission and optical density.

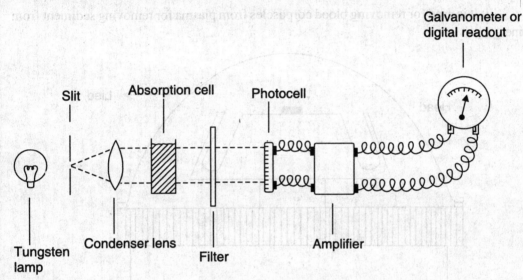

Fig. 7. Colorimeter.

When light passes through a selected filter it then passes through the sample test tube containing the solution and falls on a sensitive photocell. The current generated by this photocell is amplified by the galvanometer to indicate the percentage of light transmitted through the solution. Selection of proper filter is very essential.

USES :

(*i*) It is used for determination of the concentration of compounds in solution.

(*ii*) It can also be used to identify the compounds without isolating and purifying them.

8. CENTRIFUGE

Centrifuge is an apparatus that rotates at high speed and separates substances particles on the basis of mass, viscosity and density by means of centrifugal force and gravitational force. The centrifugal force is noted in revolution per minute (rpm) of angular speed. The force of gravity can be artificially increased in the centrifuge by increasing the Relative Centrifugal Force (R.C.F.) which depends upon the radius of the centrifuge and the revolution per minute.

COMPONENTS : Head, motor and containers. A centrifuge consists of a head which is rapidly revolved by an upright motor generally four metal cups / containers are attached to the head for holding tubes or other containers of the material from which the particulate matter is to be separated. During centrifugation, liquid containing particulate matter is kept in the tubes, runs at a particular speed and when the centrifugation is complete the particulate matter gets settled at the bottom of the tubes.

USES :

(*i*) It is used for separation of particles dispensed in a suspended manner.

(*ii*) For separation of mixture of liquids varing in their density or solids from liquids or concentrating micro-organism in various samples for enzymatic studies.

(*iii*) It is used for removing blood corpuscles from plasma for removing sediment from urine.

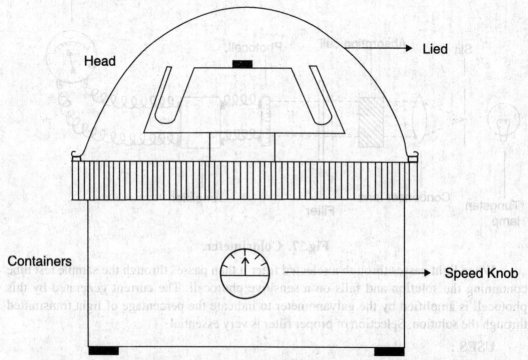

Fig. 8 Centrifuge.

PRECAUTIONS :

(*i*) Tubes must be put in pairs to balance.

(*ii*) Be sure that cotton plug if used are not forced down during the centrifugation.

(*iii*) Make sure that metal buckets are properly seated on the ring and are free to swing.

(*iv*) Bring the rheostat to zero before start, gradually increase the speed and bring the speed to required revolution per minute.

(*v*) After centrifugation, switch off and then bring the rheostat to zero position. Allow it to come to stop.

(*vi*) Hands should never be used to stop the machine.

Refrigerated Centrifuge of Large Capacity : The refrigerated centrifuge have refrigerated rotor chambers, having a speed of 6000 rpm, producing a maximum relative centrifugal field attaining 6500 g.

USES : These are used for collecting heat sensitive substances and other cells such as yeast, chloroplast, nuclei etc.

9. ULTRACENTRIFUGE

These are preparative and analytical type centrifuge having spinning rotors that can attains 8000 rpm with relative centrifugal field of up to 600,000 g.

The chamber of rotors is cooled and sealed to minimise the frictional resistance between spinning air and the rotor.

It consists of temperature monitoring unit employing an infrared, temperature sensor that can continuously monitor rotor temperature and refrigerated system.

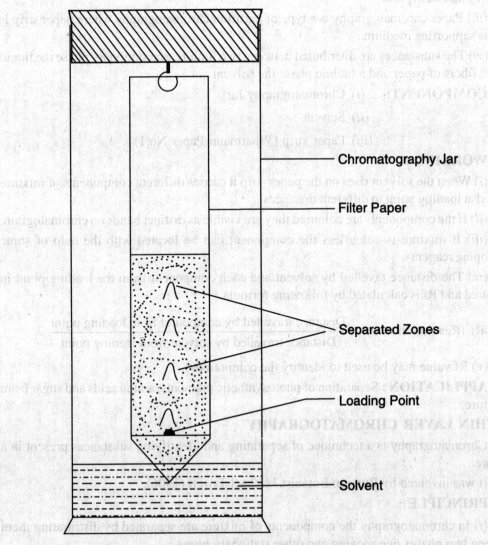

Chromatography Jar

Filter Paper

Separated Zones

Loading Point

Solvent

Fig.9. Paper Chromatography.

USES : Used for separation of small volumes of samples in biochemical and clinical laboratories for separation of major lipoprotein fractions from plasma and removal of protein from physiological fluids for amino acid analysis.

10. PAPER CHROMATOGRAPHY

(*i*) Chromatography is a technique of separating and identifying substances present in a mixture.

(*ii*) It was invented by Rusian Botanist Michael Tswett (1906).

PRINCIPLE :

(*i*) The components of a mixture are separated by distributing them between two phases one is moving and other is stationary.

(*ii*) The molecules are separated on the basis of their shape, size, mass, charges, solubility and adsorption.

(*iii*) Paper chromatography is a type of partition chromatography where paper strip is used as supporting medium.

(*vi*) The substances are distributed between two liquids, one stationary phase the liquid held in fibers of paper and a mobile phase the solvent.

COMPONENTS : (*i*) Chromatography Jar

(*ii*) Solvent

(*iii*) Paper strip (Whatmann Paper No.1)

WORKING :

(*i*) When the solvent rises on the paper strip it carries different components of mixture applied at loading point to different distances.

(*ii*) If the components are coloured they are visible as distinct bands on chromatogram.

(*iii*) If mixture is colourless the component can be located with the help of some developing reagents.

(*iv*) The distance tavelled by solvent and each component from the loading point in measured and Rf is calculated by following formula.

$$\text{Rf (Resolution front)} = \frac{\text{Distance travelled by compound from loading point}}{\text{Distance travelled by solvent from loading point}}$$

(*v*) Rf value may be used to identity the compounds.

APPLICATION : Separation of photosynthetic pigments, amino acids and sugar from a mixture.

11. THIN LAYER CHROMATOGRAPHY

Chromatography is a technique of separating and identifying substances present in a mixture.

It was invented by Russian Botanist Michael Tswett (1906).

PRINCIPLE :

(*i*) In chromatography the components of mixture are separated by distributing them between two phases one moving and other stationery phase.

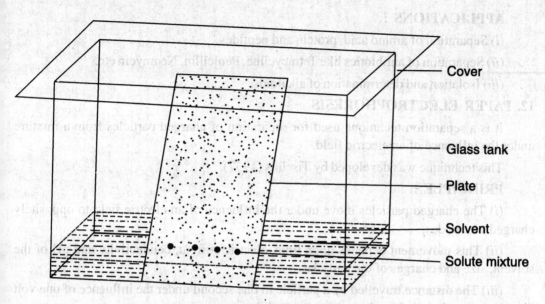

Fig. 10. Thin Layer Chromatography.

(*ii*) The molecules are separated on the basis of their shape, size mass, charges, solubility and adsorption.

(*iii*) TLC is a type of adsorption chromatography where components of a mixture are separated over a thin layer of adsorbent supported on a glass plate.

(*iv*) It is a solid liquid chromatography. Adsorbent acts as stationary phase and solvent as mobile phase.

COMPONENTS : (*i*) Glass plate coated with adsorbent (silica gel or aluminium oxide).

(*ii*) Glass Tank.

(*iii*) Solvent.

WORKING :

(*i*) The solvent rises upward on the layer of adsorbent by capillary action.

(*ii*) It also carries the components of mixture to different distances from the loading point.

(*iii*) The coloured components are visible as distinct bands on chromatogram.

(*iv*) The colourless components are located with the help of developing solvent or UV light.

(*v*) The distance travelled by solvent and each component from the loading point is measured and Rf is calculated by following formula.

$$\text{Rf (Resolution front)} = \frac{\text{Distance travelled by compound from loading point}}{\text{Distance travelled by solvent from loading point}}$$

(*vi*) Rf value may be used to identity the compounds.

APPLICATIONS :

(*i*) Separation of amino acid, protein and peptides.

(*ii*) Separation of antibiotics like Tetracycline, Penicillin, Neomycin etc.

(*iii*) Isolation and determination of alkaloids.

12. PAPER ELECTROPHORESIS

It is a separation technique used for separation of charged particles from a mixture under the influence of an electric field.

This technique was developed by Tiselius (1937).

PRINCIPLE :

(*i*) The charged particles move under the influence of an electric field to oppositely charged electrodes.

(*ii*) This movement depends upon time, electric current, conductivity and pH of the solvent, size and charges of the particles.

(*iii*) The distance travelled by a particle in one second under the influence of one volt electric current is called its electrophoretic mobility.

(*iv*) As particles of different size have different electrophoretic mobility under the influence of same current, this makes possible to separate them from a mixture.

COMPONENTS :

(*i*) Two electrode tanks containing buffer solution.

(*ii*) Supporting medium - Paper strip.

(*iii*) D.C. source.

WORKING : When mixture in placed on supporting medium and electrodes are connected with D.C. source the components of the mixture are separated in the form of distinct zones which can be located by staining with dyes.

APPLICATIONS :

(*i*) Separation of isozymes.

(*ii*) Separation of components of protein mixture.

(*iii*) In study of clearage products of nucleic acids.

(*iv*) In study of serum proteins.

13. GEL ELECTROPHORESIS

It is a separation technique used for separation of charged particles from a mixture under the influence of an electric current.

The technique was developed by Tiselius (1937).

PRINCIPLE :

(*i*) The charged particles move under the influence of an electric field to oppositely charged electrodes.

(*ii*) This movement depends upon time, electric current, conductivity and pH of the solvent, size and charges of the particles.

(*iii*) The distance travelled by a particle in one second under the influence of one volt electric current is called its electrophoretic mobility.

(*iv*) As particles of different size have different electrophoretic mobility under the influence of same current, this makes possible to separate them from a mixture.

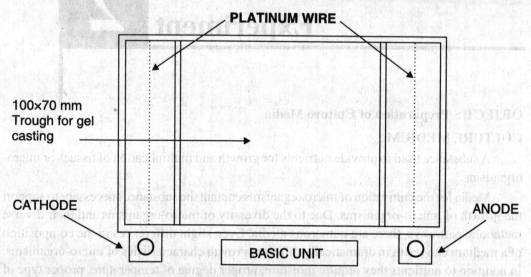

Fig. 11. Gel Electrophoresis.

COMPONENTS :

(*i*) Two electrode tanks containing buffer solution.

(*ii*) Supporting medium – Polyacrylamide gel or starch gel.

(*iii*) D. C. source.

WORKING : When mixture in placed on supporting medium and electrodes are connected with D.C. source the components of the mixture are separated in the form of distinct zones which can be located by staining with dyes.

APPLICATIONS :

(*i*) Separation of isozymes.

(*ii*) Separation of components of protein mixture.

(*iii*) In study of clearage products of nucleic acids.

(*iv*) In study of serum proteins.

Experiment 2

OBJECT : Preparation of Culture Media

CULTURE MEDIUM

A substance used to provide nutrients for growth and multiplication of tissues or micro-organisms.

Media for the cultivation of micro organisms contain the substances necessary to support the growth of micro-organisms. Due to the diversity of micro-organisms and their diverse metabolic pathways, there are numerous media. Even slight differences in the composition of a medium can result in dramatically different growth characteristics of micro-organisms. In addition to nutrients they require moisture, proper degree of temperature, proper type of gaseous environment and a suitable pH.

Media for cultivation of micro-organisms have a source of carbon for incorporation into biomass. For autotrophs, the carbon source most often is carbon dioxide, which may be supplied as bicarbonate within the medium. Carbohydrates, such as acetate, various lipids, proteins, hydrocarbons and other organic compound are included in media as sources of carbon for heterotrophs. These carbon sources may also serve as the supply of energy. Other compounds such as ammonium ions, nitrite ions, elemental sulphure and reduced iron may be used as the sources of energy. Nitrogen also is required for microbial growth. It may be supplied as inorganic nitrogen compounds for cultivation of some micro-organisms but commonly supplied as proteins, peptones or amino acids. Phosphates and metals such as magnesium and iron are also necessary components of microbiological media. Phosphates may also serve as buffer to maintain the pH of the medium within the growth tolerance limits of the micro-organism being cultivated. Various additional growth factors may also be included in the media.

Culture media are mainly of two types –

(*i*) **Liquid Media**

(*ii*) **Solid Media**

Liquid media consists of all the essential requirement of the culture medium except solidifying agents. This liquid medium can be converted to solid medium by adding suitable gelling / solidifying agents like Gelatin, Agar, Agarose etc. Most preferred is the agar powder.

16

CRITERIA FOR MEDIA PREPARATION :

(*i*) Dissolve all the ingredients in a specified amount of the distilled water. If any of the ingredient is heat labile then it should be added after sterilization.

(*ii*) Adjust the pH of medium as desired.

(*iii*) When medium is to be kept liquid, then the mixture is equally distributed into container (test tubes / flasks) and then plugged with cotton plug. In the preparation of solid media, the medium is heated and stirred slowly to dissolve the agar, when about half digestion occurs, then the medium is distributed to various containers and then plugged with cotton.

(*iv*) A paper is wrapped around the mouth of container and then the medium is sterilized.

(*v*) Both can be directly used for inoculation while the solid medium is distributed equally and aseptically as per requirement into sterile containers.

(*vi*) For solid media we can either prepare slant or stab.

1.   Nutrient Agar

   Compos\ition Per Litre

Agar	–	15.0gm
Peptone	–	5.0gm
NaCl	–	5.0gm
Yeast Extract	–	2.0gm
Beef Extract	–	1.0gm
PH		7.4 ± 0.2 at 25°C

2   Nutrient Broth

Peptone	–	5.0gm
NaCl	–	5.0gm
Yeast Extract	–	2.0gm
Beef Extract	–	1.0gm

PREPARATION OF MEDIUM : Add components to distilled / deionized water and bring volume to 1.0 liter. Mix thoroughly, Genetly heat and bring to boiling. Distribute into tubes or flasks. Autoclave for 15 min. at 15 psi pressure 121°C. Pour into sterile Petri dishes or leave in tubes.

USES : For cultivation and maintenance of a wide variety of micro-organisms.

Luria Bertani Broth

Trytone	10 gm.
Yeast Extract	5gm
Sodium Chloride	10gm
Distilled Water	Made to 1000ml

Composition Per Litre

Pancreatic digest of Casein	–	10.0gm.
NaCl	–	5.0gm
Yeast Extract	–	5.0gm

Glucose   –   1.0gm

pH   –   7.0 ± 0.2 at 25°C

4.   **L.B.Agar –**
      Composition per litre –
      Same as L.B. broth
      Agar – 15.0gm
      Use : For cultivation of E.coli

5.   **Eosin Methylene Blue Agar (E.M.B.Agar)**
      Composition Per Litre

      Agar   –   13.5gm
      Pancreatic digest of Casein   –   10.0gm.
      Lactose   –   5.0gm
      Sucrose   –   5.0gm
      K_2HPO_4 –   2.0gm
      Eosin Y   –   0.4gm
      Methylene Blue   –   0.065gm
      pH   –   7.2 ± 0.2 at 25°C
      Preparation of Medium   –   **Same**

USES : For isolation, cultivation and differentiation of gram negative enteric bacter
based on lactose fermentation. Bacteria that ferment lactose especially the coliform bacter
Escherichia coli appear as colonies with a green metallic sheen or blue black to brow
colur. Bacteria that do not ferment lactose appear as colurless or transparent. Light purp
colonies.

      MacConkey Agar
      Composition Per Litre

      Peptone   –   20.0gm
      Agar   –   12.0gm
      Lactose   –   10.0gm
      Bite Salt   –   5.0gm
      NaCl   –   5.0gm
      Neutral Red   –   0.075g.
      pH   –   7.2 ± 0.2 at 25°C
      Preparation of Medium   –   Same

USES : For the selective isolation, cultivation and differentiation of coliform a
enteric pathogens based on the ability to ferment lactose. Lactose fermenting organism
appears as red to pink color colonies. Lactose non-fermenting organisms appear as colorl
or transparent colonies.

MacConkey Broth
      All ingredients of MacConkey agar, except agar powder. pH 7.4

OBJECT – Preparation of Standard Solution.

A standard solution is that whose strength is known *i.e.,* we know the amount of substance dissolved in known volume. In volumetric analysis standard solutions are used to determine the strength of unknown solution. Some substances occur in pure state and whose composition does not change with time therefore, their standard solution can be prepared by dissolving fixed amount of substance in known volume. Such substances are known as primary standard and their standard solution can be prepared by direct weighing and then dissolving in volume of solvent (water).

1. Percentage Solution

A known quantity of solute is added to a definite volume of solution *i.e.* 100 ml expressed by 2 parameters

(*a*) Weight / volume : solute is solid here. It contains 'n' grams of solute in 100 ml of solution. *e.g.,* 1% W/V 1gm. Dissolved in 100 ml (or make up volume = 100 ml).

(*b*) Volume / volume – Solute is liquid here. It contains 'n' ml of liquid to make up volume 100 ml by adding solvent.

2. Molar Solution (M)

1 mole of a substance (solute) dissolved in solvent to make up volume – 1000 ml (1 mole = mole wt. of substance expressed in gms).

3. Molal Solution (m)

The molal of a solution is given by 1000 N where, N is the number of moles of a solute per unit mass (gm) of solvent.

4. Normal Solution (N)

A normal solution contains 1 gram equivalent dissolved in a litre of solution.

STOCK REAGENTS

DNA EXTRACTING BUFFER

	Components	Stock Conc.	Final Conc.	5ml	10ml
1	Tris (pH 8.0)	1M	50mM	0.25ml	0.5ml
2.	EDTA (pH 8.)	0.5M	25mM	0.25ml	0-5ml
3.	NaCl	5M	300mM	0.3ml	0-6ml
4.	Sodium Dodecyl Sulfate (SDS)	5M	300mM	0.3ml	0.6ml
5	Water	-	-	3.9ml	7.8ml

2.   TE BUFFER (pH 8.0)

Components	Stock Conc.	Final Conc.	5ml	100ml
1. Tris (pH 8.0)	1M	10mM	0.5ml	1.0ml
2. EDTA (pH 8.)	0.5M	1mM	0.1ml	0.2ml
3. Water	-	-	49.4ml	98.8ml

3.   STOCK

1M Tris HCl (pH 8.0) : Dissolve 30.28gm of Trizma base in 200ml of distilled water Adjust the pH to 8.0 with concentrate. HCl (about 10.5ml) allow the solution to cool to room temperature 25°C before making final adjustment of the pH. Adjust the volume of the solution to a total of 250ml with distilled water. Sterilize by autoclaving.

0.5M EDTA (pH 8.0) : Add 46.53 gm of Disodium Ethylenediam Inetetraacetate (EDTA) with $2H_2O$ (sigma FW=372.2) to 200ml of distilled water stir vigorously with magnetic stirrer with 4gm. NaOH. Adjust the total volume to 250ml with distilled water Sterilize by autoclaving.

5M NaCl : Dissolve 73.05g. (sigma, FW = 58.44) in 200ml of distilled water. Adjus the total volume to 250ml with distilled water. Sterilize by autoclaving.

1M Tris (pH 8.4 at 25°C) : Dissolve 30.28 gm. of Trizma base (sigma) in 200ml o distilled water. Adjust the pH to 8.4 with concentrate HCl (about 10ml). Allow the solutio to cool to room temperature (25°C) before making a final adjustment of the pH. Adjust th volume of the solution to 250ml with distilled water. Sterilize by autoclaving.

1M KCl : Dissolve 18.64 gm of potassium chloride in 200ml of distilled water an sterilize by autoclaving.

10X Loading Buffer : Mix 40 mg. of bromophenol blue (final 0.4%) 40mg of xyle cyanole (final 0.4%) and 5ml glycerol (final 50%). Adjust the final volume to 10ml wi distilled water. Aliquot to 1.5ml micro centrifuge tubes, heat in boiling water for 10 mi cool down and store at 4°C.

50X TAE : Mix 24.2 gm of Trizma base, 57.1ml a glacial acetic acid and 100ml 0.5M EDTA (pH 8.0) with 700ml distilled water. Adjust the final volume to 1000ml.

Experiment 4

OBJECT : Isolation of bacteria by Streak plate method.

REQUIREMENTS : Nichrome wire loop, bunsen burner, nutrient agar plate, culture, glass marker.

PRINCIPLE : Generally bacteria exist in mixed population. Isolation means to separate the micro-organisms from mixed population. For studying the cultural and morphological characters of an individual species, it is essential to separate them from the other species to get in the form called pure culture. By means of a transfer loop, a portion of the mixed culture is placed on the surface of an agar medium and streked across the surface. Streak plate method "thins out" the bacteria on the agar surface so that some individual bacteria are separated from each other. When streaking is properly performed, the bacterial cells will be sufficiently for apart in some areas of the plate to ensure that the colony developing from one cell will not merge with that growing from another.

METHOD :

(*i*) Take a nutrient agar plate and make 4 sectors by glass marker on the back side of the plate.

(*ii*) All the 4 sectors must have some overlapping area with the next sector (sector no 1 & 4 should not be overlap, these sectors must be as apart as possible).

(*iii*) Sterilize the wire loop and take a loopfull of culture under aseptic condition and streak out sector number 1 uniformly.

(*iv*) Sterilize the wire loop again and streak out sector number 2 without taking culture again. Organisms from overlapping area of sector 1 of 2 will be transferred to the sector 2.

(*v*) Again sterilize the wire loop and streak out sector number 3 and 4.

(*vi*) Streak out central area.

(*vii*) Incubate the plate in an inverted position at 37°C for 24 hours.

(*viii*) After incubation write down the colonial characteristics from isolated colony.

OBSERVATIONS :

Colony characters.

S. No.	Shape	Edge	Elevation	Opacity	Texture	Pigmentation

RESULT : The isolated colony of desired micro – organisms on the plate will be observed.

PRECAUTIONS :

(*i*) Streaking must be properly performed.

(*ii*) All the 4 sectors must have some overlapping area with the next sector (but sector number 1 and 4 should not be overlap).

(*iii*) Take a culture only in one time.

(*iv*) Plates are incubated in an inverted position.

(*v*) The whole experiment is performed under aseptic condition.

Experiment 5

OBJECT : Isolation of bacteria by pour plate method.

REQUIREMENTS : Melted nutrient agar tubes, strile petri plates, culture suspension, nichrome wire loop, bunsen burner.

PRINCIPLE : The pour plate method is based on diluting the mixed culture with liquefied nutrient agar in such a manner that the colonies formed on the plates are countable. This method gives an idea of the viable bacterial count in a suspension and is used for isolation of pure culture of bacteria.

The mixed culture is first diluted to provide only a few cells per milliliter before being used to inoculate media since the number of bacteria in the specimen is not known before hand, a series of dilutions is made so that at least one of the dilutions will contain a suitably sparse concentration of cells.

METHOD :

(*i*) One loopfull or 0.1ml of original culture suspension is transferred to tube A (liquid cooled agar medium).

(*ii*) Tube A is rolled between the hands to effect through mixing of inoculum.

(*iii*) Similar transfer are made from A to B to C.

(*iv*) Contents of each tube are poured into separate sterile petri plates.

(*v*) Allow the media to solidify and incubate the plate at 37°C for 24 hours.

(*vi*) After incubation, plates are examined for the one which contains well separated colonies and note down the colonial characteristics from isolated colonies.

(*vii*) Evaluate the plates for obtaining pure culture.

OBSERVATIONS :

Colony characters.

S. No.	Shape	Edge	Elevation	Opacity	Texture	Pigmentation

23

RESULT : The isolated colonies of micro – organisms will be visible.

PRECAUTIONS :

(*i*) The nutrient agar medium is maintained in a liquid state at temperature 45 - 50°C before the pouring the plate.

(*ii*) Care should be taken while transferring the inoculum from one tube to another.

(*iii*) The whole experiment is performed under aseptic condition.

Experiment 6

AIM : To determine the growth characteristics of *E.coli* using plating technique.

REQUIREMENTS : Nutrient agar plates, Eosin Methylene Blue agar plates, MacConkey agar plates, pure culture of *E.coli*. Test tube, nicron wire loop etc.

PRINCIPLE : When grown on a variety of media, micro-organisms (*E.coli*) will exhibit differences in the macroscopic appearance of their growth. These differences, called cultural characteristics, are used as bases for separating micro-organisms into taxonomic groups.

METHOD :

(*i*) Prepare a N.agar, MacConkey agar plates and E.M.B agar plates.

(*ii*) Take a plate (N.agar) and marke 4 sectors by glass marking pencil on the back of the plate.

(*iii*) Sterilize the wire loop and take a loopful of culture under aseptic condition and streak out sector number 1 uniformly.

(*iv*) Sterilize the wire loop again and streak out sector number 2 without taking culture again. Organisms from overlapping area of sector 1 of 2 will be transferred to the sector 2.

(*v*) Sterilize the wire loop and streak out sector 3 and sector 4.

(*vi*) Streak out central area.

(*vii*) Repeat the procedure (No.2 – 7) with another plates *viz.,* MacConkey agar plates nd E.M.B. plates.

(*viii*) Incubate the plates in an inverted position at 37°C for 24 hrs.

(*xi*) After incubate write down the colonial characteristis.

OBSERVATIONS : Colony characters

S. No.	Media Used	Shape	Edge	Elevation	Opacity	Texture/ Pigmentation
(*i*)	N. agar					
(*ii*)	MacConkey Agar					
(*iii*)	E.M.B. Agar					

On MacConey agar the colonies are bright pink because of lactose, fermentation. On EMB agar colonies with dark center and greenish metallic sheen.

RESULT : The growth characteristics of *E.coli.* are ....

PRECAUTIONS :

(*i*) Experiment should be done strictly under aseptic conditions.

(*ii*) Streaking should be done carefully.

(*iii*) Prepare the N. agar, MacConkey agar and E.M.B. agar plate carefully.

Experiment 7

AIM : Growth characteristics of *E.coli* using turbidimetric method.

REQUIREMENTS : N. agar broth, N. agar plate pure culture of *E.coli* shaker, calourimeter, conical flask, incubator, nicron wire loop.

PRINCIPLE : It is a method of estimation of bacterial cell growth or population in broth, culture or aqueous suspension by the measurement of the degree of opacity or turbidity of the suspension. Here, the growth is determine by photometric method in terms of optical density where the basic principle is that the optical density is directly proportional to the micro – organism present.

METHOD : Prepare a N. agar borth and N. agar plate.

(*i*) Sterilize the wire loop and take a loopful of *E.coli* culture under aspectic condition and transfer it to conical flask containing N. broth and place it in shaker for 30 min.

(*ii*) Adjust the colourimeter and glass tube (cuvelte) filled with uninoculated culture medium is then used to set the instrument to give a basal optical density reading of 0.

(*iii*) The blank tube is replaced by a similar tube containing the broth culture (*i.e.* – medium + cells) and the increase in optical density is recorded.

(*iv*) Repeat the same procedure for 24 hrs. and record the O.D. every one hr.

OBSERVATION :

Aliquots	Time in Hrs.	Optical Density	Turbidity Index*
1			
2			
3			
4			
5			

RESULTS : Optical density is directly proportional to the micro – organisms present a broth culture.

Turbidity Index

(*i*) Maximum turbidity (Max. growth) +++

(*ii*) More turbidity (More growth) ++

(*iii*) Low turbidity (Least growth) +

(*iv*) No turbidity (No growth) -

Plot a graph between time and optical density to get growth curve.

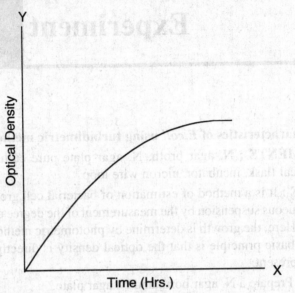

PRECAUTIONS :

(*i*) Measure optical density of bacterial sample by taking absorbance at 610nm.

(*ii*) Always set the instrument before taking OD to zero with uninoculated sample.

(*iii*) Prepare the N. broth and N. agar plate carefully.

Experiment 8

OBJECT : Isolation of plasmid from _E.coli_ by alkaline lysis method and to analyze plasmid DNA by agarose gel electrophoresis.

MATERIALS REQUIRED :

Equipments : Incubator, shaker, microcentrifuge electrophoretic unit, Vortex mixer, microwave / heater, laminar air flow.

Glassware : Beaker, conical flask, test tubes, measuring cylinder, peteriplate, staining tray. Reagent and other bacterial host plate (containing plasmid PUC 18), solution I, II & III, solution IV (Isopropanol), Ampicillin, Gel loading dye, R Nase, Agarose, 50XTAE, XTE, 70% ethanol L.B. broth, L.B. plates, 6Xstaining dye, control DNA, autoclaved distilled water .

Other Requirements : Crushed ice, micropipettes, sterile tips, blotting paper, 1.5ml ials.

Solution – I	15% glucose	–	Prevents immediate lysis of cell
	25mM tris	–	Maintains pH
	10mM EDTA	–	Chelates metal ions and weakens cell wall
	100 µg/µl RNAse		
Solution–II	0-2N NaOH	–	Denature chromosomal & plasmid DNA
	1% SDS	–	Denatures bacterial proteins & disrupts cell membrane
Solution-III	3M Sodium Acetate	–	Renatures plasmid DNA
Solution-IV	Isopropanol	–	Precipitates plasmid DNA
RNAse	Degrades RNA without affecting DNA		

PRINCIPLE : Plasmids are double stranded, circular, self replicating extra-chromosomal DNA molecules. They are commonly used as coloning vectors in molecular biology survive in non-ideal condition.

Alkaline lysis method for rapid purification of plasmids exploits the topological difference between plasmid circles and linear chromosomal fragments. Cells are lysed by treating with alkali (NaOH) and a detergent, sodium dodecyl sulphate (SDS). SDS denatures bacterial proteins and NaOH denatures the plasmid and chromosomal DNA. However, in case of plasmids the strands remain closely circularized since they are linked by the

29

interwined backbones of double helix. In contrast strands of linear / nicked DNA are free to separate completely. When this mixture of denatured plasmid and chromosomal DNA is neutralized by the addition of sodium acetate, renaturation occurs. Renaturation of plasmid is rapid and accurate since the strands are already in close physical proximity. Linear molecules generated by random shearing of chromosomal DNA renature less accurately forming networks of DNA that can be removed from lysate by centrifugation, together with denatured protein and RNA. Plasmid DNA remains in solution and can be precipitated using ethanol / isopropanol. Agarose gel electrophoresis separates plasmid DNA by its size and shape.

PROCEDURE :

(i) **Day–I** Streak the given bacterial culture on L.B. Agar containing ampicillin (100 mg/ml) .

(ii) Incubate at 37°C overnight.

(iii) **Day-II** Pick a single colony from the L.B. plate and inoculate into 10ml L.B. broth containing ampicillin (100 µg/µl). Incubate at 37°C with shaking for 8 – 16 hrs.

(iv) **Day-III** Pipette 1-5 ml culture into 1-5ml vial.

(v) Spin at 6000 rpm for 8-10 minutes Discards the supernatant and invert the vial or blotting paper to drains out left over supernatant, place on ice.

(vi) Resuspend cell pellet in 100 µl of ice cold solution-I and mix the contents by finger flicking the tube. No visible clumps of bacteria should remain. Place on ice for 5 minutes and shift to room temperature (RT).

(vii) Add 200 µl of solution-II at RT. Mix gently by inverting the vial five times. The cell suspen sion should look clear at this stage.

(viii) Add 150 µl of solution-III. Mix genetly by inverting the vial. Place on ice for 10 min.

(ix) Spin at 6000 rpm for 30 minutes.

(x) Transfer the supernatant immediately to a fresh vial and add 450 µl of solution-IV to precipitate the DNA. Mix by inverting the vial. Incubate at RT for 10-15 minutes.

(xi) Spin at 6000 rpm for 30 minutes. Decant the supernatant. Invert the vial on blotting paper to drain out left over supernatant DNA will be seen as white precipitate, sticking to the side of the vial.

(xii) Add 200 µl of 70% ethanol to opposite side of the wall from the pellet Drain off the alcohol and keep the tube inverted on paper towel.

(xiii) When the pellet turns transparent, add 50 ml of IXTE to the pellet and resuspend by finger flicking.

(xvi) Add 5 µl of RNAse to vial and keep it for 30 min. label these vial as RNAse treated.

(xv) Meanwhile, prepare 1% agarose gel.

(xvi) Prepare 50 XTAE buffer (50 XTAE – 4ml TAE + 200ml D/W)

(xvii) Add 2 µl of gel loading buffer to each of the sample.

(xviii) Load 10-15 ml of your DNA sample after addition of gel loading dye and also load control DNA sample on 1% agarose gel at 50V for 1-2 hours.

(xix) Visualize with UV doc system or stain with 6Xdye (staining dye) and note the observations.

OBSERVATION : Under UV light, DNA fluoresces. DNA is visible as a dark blue band against a light blue background. Compare the extracted DNA samples with that of control DNA and observe for bands corresponding to supercoiled or niked / linear forms. Also observe for presence or absence of RNA in samples untreated or treated with RNAse.

PRECAUTIONS :

(*i*) Streaking and inoculation should be done aseptically.

(*ii*) Pipette should be handled carefully.

(*iii*) Care should be taken for preparing 1% agarose gel and TAE buffer.

(*iv*) Carefully load the samples in the well on agarose gel.

Staining Procedure to Visualize DNA :

(*i*) Prepare 1X staining dye by diluting 6Xdye (1:6) (Bromophenol blue) with distilled water (Approximately 50ml of 1X staining dye is required for one experiment. Therefore, make up 8ml of 6Xdye to 48ml with distilled water).

(*ii*) Carefully transfer the gel (from gel tank) into a tray containing IX staining solution. Make sure that the gel is completely immersed.

(*iii*) For uniform staining, place the tray on a rocker for approximately one hour or shake intermittently every 10 to 15 minutes.

(*iv*) Pour out the staining dye into a container distain the gel by washing with tap water several times till the DNA is visible as a dark band against a light blue background.

OR

(*i*) Alternatively Ethidium bromide can be used for visualizing DNA fragments. Add Ethidium bromide to motten agarose to a final conc. of 0.5 mg/ml (from a stock of 10 mg/ml in water).

(*ii*) When temperature is around 50ºC, mix and cast the gel. After electrophoresis, DNA samples can be visualized under UV light. They appear fluorescent.

INTERPRETATIONS :

(*i*) On analyzing plasmid DNA after electrophoresis, one observes two bands corresponding to nicked and super coiled DNA. As is seen the super coiled form runs faster than nicked form due to its compact structure.

(*ii*) RNA, which is a small molecule, is seen migrating faster than the super coiled DNA in samples not treated with RNAse.

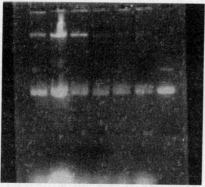

Plasmid Isolation

PLASMID ISOLATION FLOW CHART

Day-I   Preparation of Media and Revival of Stain
                        ↓

Day-II   Inoculation of Media
                        ↓

Day-III   Spin 1.5ml Culture 6000 rpm 5 min.
                        ↓ Drain

            Pellet + Solution-I 100 µl
                        ↓ 5 min. R.T

            Pellet + Solution-II 200 µl
                        ↓

            Pellet + Solution-III 150 µl
                        ↓

            Spin 10000 rpm min.
                        ↓   Decant Pellat

        Supernatant + 0.5 ml Isoproponal
                        ↓

            10000 rpm Spin
                        ↓

            70% ethanol Wash
                        ↓

            Air dry
                        ↓

            +50 µl TE (1X)
                        ↓

        Load on to 1% agarose gel

Experiment 9

OBJECT : Isolation of bacterial genomic DNA from *E.coli* or Serratia marsences.

MATERIAL REQUIRED : Cell pellets of *E.coli* or *Serratia marsences* in 1.5ml eppendorff vials, cell lysis buffer, DNA rehydrating solution, control DNA, Gel loading dye and buffer. 50 x TAE, Agarose, ethanol 70%, 1.5ml vials, Microcentrifuge, Micropipettes, tips, Transilluminator, staining tray, distilled water, Beaker.

PRINCIPLE : Cells are first broken and then DNA is separated from other components such as proteins, RNA, Lipids and carbohydrates. Cell lysis solution contains 4M guanidium thiocyanate salt as effective protein denaturant and a strong inhibitor of ribonuclease and deoxyribonucleases. Cell lysis occurs due to the action of guanidium thiocyanate and the detergent sarkosyl present in the cell lysis buffer. Upon centrifugation cell debris along with the trapped RNA and proteins are separated. The resulting supernatant mainly consists of genomic DNA and a slight amount of RNA. The nucleic acid is then precipitated using alcohol.

PROCEDURE :

(*i*) Remove the vial containing the bacterial cell pellet from ice and thaw at room temperature.

(*ii*) Re-suspend the cells in 700µl of cell lysis solution at room temperature.

(*iii*) Incubate at room temperature for 5 minutes and spin at 10000 rpm for 10 minutes.

(*iv*) Collect 500µl of the supernatant in a fresh tube (a jelly like pellet of cell debris is seen. Avoid decanting this pellet).

(*v*) To the 500µl of the supernatant add 1ml alcohol / isopropanol. Mix by inversion till you see white strands of DNA precipitating out.

(*vi*) Spool this DNA with the help of a tip and transfer into a fresh tube or spin at 12000 rpm for 5 minutes and discard the supernatant.

(*vii*) Wash the DNA pellet with 5ml of 70% ethanol and give a short spin at 10,000 rpm, decanting it and repeat this step. Give a final wash with 75% ethanol and air dry for 5 minutes.

(*viii*) Add 100µl of DNA rehydrating solution B and incubate at 55-60°C for 5 minutes of increase the solubility of genomic DNA.

(*ix*) To get rid of insoluble material spin at 12,000 rpm for 10 minutes and pipette out the supernatant into a fresh tube.

(*x*) Take 30µl of the freshly isolated DNA along with 5µl of gel loading dye, mix and load into the gel. Take 10µl of control DNA and electrophorise along with isolated samples on 1% agarose gel (note down the order in which the sample have been loaded).

(*xi*) Electrophorese the sample at 100 volts for 2-3 hours.

(*xii*) Stain with IX staining dye (Bromophenol blue).

(*xiii*) Destain to visualize the band(s).

OBSERVATIONS : For visualizing DNA, cut the gel, lift and place on the transilluminator DNA can be seen as orange band under uv (if ethidium bromide is used)

OR

The gel is observed against a light background wherein DNA appears as dark coloured bands (if bromophenol blue is used). Compare the mobility of extracted DNA with that of control DNA.

RESULTS AND INTERPRETATION : The molecular weight of control DNA band is around 50 Kb in size. From the gel, one can observed the Genomic DNA being high mole wt. (equal or above 50 Kb), should run along with the control DNA or above if shearing has not occurred during isolation, one may see DNA band below the control DNA. If RNA is present along with the isolated DNA, it will be seen between the blue and purple dye or on the purple dye. It is migrating faster as it is a smaller molecule compared to DNA.

PRECAUTIONS :

(*i*) Handle solution A with care as it is corrosive in nature.

(*ii*) Prepare required amount of 70% alcohol using distilled water, just before use.

(*iii*) Wear gloves while handling ethidium bromide stained.

Experiment 10

OBJECT : To determine the DNA molecular size.

MATERIAL REQUIRED

Equipment : Electrophoresis unit.

Glassware : Beaker, Conical Flask, Measuring Cylinder, Staining Tray.

Reagents and Other Requirements : Agarose, Gel loading buffer (3X) (GLB), staining dye (6X), 50XTAE, Molecular weight marker 500bp DNA. Ladder, test sample I, II & III, Micropipettes, 1.5ml Vials, sterile distilled water.

PRINCIPLE : Linear DNA molecular size is determined by comparing the electrophoretic mobility of the unknown DNA with the standard DNA on agarose gel.

In this method, standard molecular weight DNA marker and unknown sample DNA molecules are run on agarose gel. The distances migrated by standard DNA molecule and sample DNA molecules are measured and plotted on the graph. The size of unknown DNA molecule is then determined from graph.

PROCEDURE :

(*i*) Preparation of 1% Agarose gel and setting up of electrophoresis.

(*a*) Prepare 1XTAE by diluting appropriate amount of 50XTAE buffer with distilled water (4 ml/TAE + 200 ml D/W).

(*b*) Take 50ml of 1XTAE in a 250ml conical flask and add 0.5g. of agarose. Boil to dissolve agarose (till a clear solution results) cool to a warm liquid.

(*c*) Place the combs of the electrophoresis set such that the comb is about 2cm. away from cathod.

(*d*) When the agarose gel temperature is around 60°C, pour the cooled agarose solution slowly into the gel tank. Make sure that the agarose gel is poured only in the center part of the gel tank and is 0.5 to 0.9 cm. thick, without air bubbles. Keep the set undisturbed till the agarose solidifies.

(*e*) Once, the gel has solidified, pour 1XTAE buffer slowly into the gel till the buffer level stands 0.5 – 0.8 cm above the gel surface.

(*f*) Remove the combs gently so that the wells are intact.

(*g*) Connect the power cord, the red cord to the red electrode and the black cord to the black electrode. Do not switch on the power before loading.

(h) Load the samples into the well. Record which samples are being loaded into which wells as Lane 1, Lane 2...... After loading start the power connection. Set the voltage 50V/100V.

(i) Run till the second dye (blue dye) from the well has reached 3/4th of the gel (1 hr. approx) use the staining dye (Bromophonol blue) provided to you for staining after electrophoresis. Dilute the staining dye 1:6 with distilled water before using.

(ii) Take 1 μg of molecular size marker (500 bp ladder), test samples 1,2,3 in sterile 1.5ml vials, add 3 μl of GLB mixwell and load the samples onto the gel.

(iii) Run the gel at 50V for 1-2 hrs. until dye reaches the bottom of agarose matrix. (Do not let the dye run out).

(iv) Staining of Gel – Remove the gel by slowly running a spatula along the walls of the gel tank. Place the gel block in a small tray and pour the staining dye on it. Make sure that the gel is completely immersed and shake the tray slowly. Staining takes approximately half or one hr. shake the tray every 5 to 10 minutes. Remove the staining dye in a container. Destain the gel by washing with tap water several times, till the DNA is visible as a dark band against a light blue background.

(v) Measure the distance traveled by the dye, 500 bp ladder bands and test samples 1,2 & 3 from the well in cm and tabulate as below -

(vi) Calculate the Rf value using the formula

$$Rf = \frac{\text{Distance travelled by DNA Molecule}}{\text{Distance Travelled by the Dye}}$$

(vii) Construct a graph by plotting the Rf values against the corresponding molecular size of the ladder.

(viii) Calculate Rf value for sample DNA and obtain the corresponding molecular size from the graph.

OBSERVATIONS : DNA is visible as a dark band against a light blue background.

Size of Test Sample	1	_____kb (base pair)
	2	_____kb
	3	_____kb
	4	_____kb

PRECAUTIONS :
(i) Pipette should be handled carefully.
(ii) Care should be taken for preparing 1% agarose gel and TAE buffer.
(iii) Carefully load the samples in the well on 1% agarose gel.

Experiment 11

OBJECT : Purification of DNA from agarose gel using silica.

MATERIAL REQUIRED

Silica Solution, 6M NaI solution; wash buffer, 1XTE, Agarose, 1XTAE, Ethidium bormide, PUC / Taq I Digest.

Instruments : Micropipette, Vortex Machine, Electrophoresis tank, combs and chords, power pack, table top centrifuge, Dry bath, 1.5ml eppendorf vials.

PRINCIPLE : Here agarose gel is used to separate DNA fragments from 50 bp to several thousand bases in length. Migration of DNA depends upon the size and conformation of the DNA. When gel slice of interest is dissolved in NaI, a chaotropic salt, that at conc. of around 4M is able to solubilize agarose. Glass beads are then added, which in this concentration of NaI efficiently bind the released DNA fragments. RNA, proteins and other impurities do not bind to the glass beads. Following a few washing and pelleting cycles the purified DNA is eluted from glass into a low salt buffer.

METHOD :

(*i*) Prepare 1% agarose gel. Add 2 ml of 10 mg/ml ethidium bromide to this hot solution and mix gently without generating air bubbles.

(*ii*) Pour 1% agarose gel into gel tank with comb.

(*iii*) After the gel solidifies, pour 1X TAE buffer till the gel is submerged and remove the comb. connect the gel tank to the power pack.

(*iv*) Load 20 μl of predigested PUC 18/Taq1 digest into the well.

(*v*) Electrophorese at 50V for 2-3 hrs.

(*vi*) Visualize the gel under UV light and cut out the DNA fragment precisely with sharp blade. Take care to cut out minimum gel slice if possible.

(*vii*) Weight the gel slice by placing in a 1.5ml vial, 1 gm. of gel is approx. equal to 1ml (volume).

(*viii*) Add 2.5 volume of 6M NaI solution and melt the gel slice by placing in 55°C dry bath. Keep mixing during incubation intermittently. Gel should melt in 10-12 minutes.

(*ix*) Vortex till it form a homogenous mixture solution.

(*x*) Add 15 μl of silica solution to this tube. Mix gently and incubate at room temperature for 10-15 minutes mix at every 1-2 minute interval. Longer incubation period facilitate greater binding of DNA.

(*xi*) Spin at 12000 rpm for 1 minute and discard the supernatant.

(*xii*) Resuspend the pellet in 200 μl of wash buffer thoroughly by vartexing.

(*xiii*) Spin at 12000 rpm for 1 min. and discard the supernatant.

(*xiv*) Repeat step 12 and 13 twice.

(*xv*) Remove the wash buffer. Traces of wash buffer should be removed by incubating for 2-5 minutes at 37°C to allow all the wash buffer to evaporate.

(*xvi*) Elute the DNA by adding 20 μl of 1XTE to the silica pellet. Resuspend thoroughly and incubate at 55°C for 5 minutes.

(*xvii*) Spin at 12000 rpm for 1 min. transfer the supernatant to a fresh 1.5ml vial.

(*xviii*) Repeat the step 16 and 17. Pool the supernatants.

(*xix*) Remove traces of silica from eluted DNA by spinning at 12000 rpm for 1 min and removing the supernatant to a fresh vial.

(*xx*) Load the purified DNA mixed with 2 ml of gel loading buffer along with the original sample of PUC 18/Taq1 digest on 1% agarose gel and electrophorese.

(*xxi*) Compare the intensity of the DNA band purified with that of the original sample

RESULT : On analyzing DNA after electrophoresis. We observe supercoiled DNA as a single band.

PRECAUTIONS :

(*i*) Be careful while preparing 1% agarose gel not to over incubate it.

(*ii*) Silica sediments very fast, DNA does not bind properly to sedimented silica. Mix thoroughly every 2-3 minutes during incubation.

(*iii*) Excess of alcohol/isopropanol from wash buffer can lead to low yields of DNA be sure to remove last traces of wash buffer by incubating at 37°C for few minutes.

(*iv*) Do not dry the pellet in vacuum drier as too dry pellet may lead to lower yields.

(*v*) Do not use glass tubes to place the gel slice as DNA sticks to glass.

(*vi*) The purification should be ideally done on the same day as electrophoresis.

(*vii*) Overloading of DNA may lead to mixing of different bands to a certain extent (if solution is very less), resulting in impure DNA preparations.

OBJECT : To perform Bacteriophage Titration.

REQUIREMENTS : Host phage lysate (*E.coli*, phage lambda), 20% Maltose, SM. Buffer, 6 tubes of dilution broth (L.B. Broth), 6 plates of L.B. agar, 6 tubes of soft agar (melted and kept at 42°C), eppendorf vials (1.5ml), spectrophotometer, centrifuge, laminar air flow, pipettes, water bath, incubator, petriplates, conical flasks, measuring cylinder, capped centrifuge tubes, distilled water, crushed ice.

PRINCIPLE : The bacteriophage titration is experiment designed to enumerate bacteriophage particles, when appropriate dilution of phage and host placed and incubate on suitable medium, the presence of phage clearly indicated by appearance of plaques on bacterial lawn. Plaque formation is result of lytic cycle adopted by bacteriophage. The phage recognizes the maltose binding protein (MBP) of *E.coli* as its receptors on the surface of the cell wall adsorbs to it. In lytic cycle a bacteriophage injects its DNA into the host, when it replicates many fold. The bacteriophage gene products are synthesized, daughter particles one assembled and the host cell eventually lyses release its many new infectious virus particles.

PROCEDURE

Day-I : Preparation of Media and Revival of Host –

1. **Bottom (Hard) Agar Plate** – It is prepared by adding 1.5gm of agar to the 100ml distilled water in a 250ml of conical flask.

   (a) Plug the conical flask containing medium with cotton.

   (b) Autoclave at 121°C, 15 lb pressure for 15 minutes.

   (c) After autoclaving remove the medium prepared and keep it for 5 min. at room temperature.

   (d) Pour the medium into sterile petriplates (15ml each plate) and allow it to solidify.

   (e) Seal the plates with parafilm or wrapped in aluminium foil.

2. **Preparation of Soft Agar** – For this add 0.6 – 0.8gm. agar to 100 ml distilled water boil it to dissolve agar and autoclave it. Aliquote 5ml. each into test tubes. Cotton plug and sterilized it and keep it at 42°C water bath.

3. **Revival of Host :** (*a*) Break open the lyophilized vial and suspend the sample by adding 0.1ml of L.B. broth.

(*b*) Streak a loopful of this suspension on to LB plate.

(*c*) Incubate the plates for 24 hrs. at 37°C.

Day-II :

(*i*) Inculate single colony from revived plate in 5ml of L.B. broth containing 0.2% maltose and 10mM $MgSO_4$.

(*ii*) Incubate it at 37°C for 24 hrs. in a shaker.

Day – III : Preparation of Plating Cells :

(*i*) Take 25ml L.B.broth to it add 0.2% maltose and 10mM $MgSO_4$.

(*ii*) Inoculate the L.B.broth with 0.25ml (1%) of overnight grown culture.

(*iii*) Incubate at 37°C for 2-3 hours and check the O.D. (O.D.-reaches 0.6 at A_{600}).

(*iv*) Chill the broth on ice for 10 minutes.

(*v*) Centrifuge the broth at 5000 rpm for 10 minutes in a sterile centrifuge tubes. Discard the supernotant.

(*vi*) Gently resuspend the cell pallet in 5-6ml of 10mM $MgSO_4$ and store at 4°C.

(*vii*) Titration - Pipette 1ml of SM buffer and pour it into each six serially labeled 1.5ml vials.

(*viii*) Take 10 µl of stock lysate and transfer into 1µl of SM buffer thus making the dilution 10^{-2} Mix and transfer 10µl of 10^{-2} dilution to vial #2 (i.e. 10^{-4} dilution) repeat the same till the dilution of 10^{-12}.

(*ix*) Label the L.B. plates. Pipette 100ml of plating cells in 6 sterile vials. Add 10ml of the respective phage dilution, mix gently and keep at 37°C for 15 minutes for adsorption of the phage onto host cell.

(*x*) Pipette out the content of vial labeled as 10^{-2} into a test tube containing 5ml of soft agar and mix between the palm. Care should be taken to ensure that the temperature of soft agar does not exceed 45°C (as the host cell will die) or fall below 40°C (as soft agar will solidify).

(*xi*) Pour the mix (phage plating cells with soft agar) immediately on the respective L.B. agar plate and allow it to solidify.

(*xii*) Repeat the step 10 and 11 for each dilution.

(*xiii*) Incubate the plates for 24 hrs. at 37°C.

OBSERVATION :

(*i*) Observe the plates for distinct plaques.

(*ii*) Count the number of phage plaques on each of your plates and record your results.

(*iii*) Determine the number of phage particle per ml of the stock solution (lysate).

Phage Titre Value = Number of Plague forming units/ml of lysate.

S.No.	Dilution	No. of Plaques	Phage Litre Value

e.g., the plate labeled 10^{-7} has 138 plaques. The titre or No. of infective particles per ml of lysate will be –

= 138×10^{-6} / $10\mu l$

= 1.3×10^{10} / μl

INTERPRETATION :

(*i*) At dilution $10^{-2} - 10^{-4}$, distinct plaques are not observered as complete lysis of the host occurs.

(*ii*) At 10^{-6} dilution, confluent plaques are observed.

(*iii*) At dilution 10^{-8} to 10^{-12}, distinct plaques are observed against the bacterial lawn due to reduction in the number of phage particles infecting the bacteria.

PRECAUTIONS :

(*i*) The experiment should be done strictly under aseptic conditions.

(*ii*) Revive the strain as soon as the lyophilized vial is opened.

(*iii*) Always use sterile distilled water.

(*iv*) Be ensure that no moisture is present on the surface of hard agar plates.

(*v*) Mix the phage lysate dilutions thoroughly by vortexing or inverting the vial.

(*vi*) While preparing dilution, change the tip after every dilution.

BACTERIOPHAGE TITRATION FLOWCHART

Day-I   Preparation of Media and Revival of Host
↓

Day-II   Inoculation
↓

Day-III   Preparation of Plating Cells 25ml
L.B.Broth + 0.25ml overnight grown culture.
↓

Incubate at 37°C for 2-3 hrs.
↓

Chill the broth on ice for 10 min.
↓

Centrifuge at 5000 rpm for10 min.

↓

Pallet + Supernatant (Discard)

↓

Pellet + 5-6ml of 10mM $MgSO_4$ Store at 4°C

↓

1ml SM buffer to each 6 1.5ml vials

↓

10µl stock lysate + 1ml of SM Buffer (dilution 10^{-2}) Mix and transfer
10ml of 10^{-2} dilution to vial 2 (10^{-4} dil.)

↓

Repeat the same till the dil. of 10^{-12}

↓

100µl plating cells in 6 sterile vials + 10µl of respective phage dilution
Mix & Keep at 37°C for 15 min.

↓

Content of vial + 5ml of soft agar Mix between palm

↓

Pour mix (Phage plating cells with soft agar) to L.B. agar plate.

↓

Repeat the Step 11 & 12 for each dilution

↓

Incubated the plates for 24 hrs. at 37°C

↓

Observe the plate for plaque formation

Experiment 13

OBJECT : **To perform restriction digestion of λ DNA with enzymes viz. Eco RI and Hind-III and to analyze the restriction pattern by agarose gel electrophoresis.**

MATERIAL REQUIRED –

Substrate DNA (λ DNA), plasmid DNA Restriction enzymes Eco RI, Hind-III, Assay buffer, Agarose, 50XTAE, Gel loading buffer, staining dye (Bromophenol blue) or Ethidium bromide, dry bath, beakers, conical flask, measuring cylinder, staining tray, distilled water, micropipette, tips, Ice, electrophoretic unit, (Gel tank, combs, cords and power supply).

PRINCIPLE : The restriction endonuclease enzymes provide a defence system in bacteria against the attack of bacteriophages. These enzymes cut phage DNA from their recognition sites and inactivate it. Bacteria protect their own DNA from these enzymes by methylating the adenine or cytosine bases which block the binding of restriction enzymes.

Each restriction endonuclease has specific recognition site on DNA is treated with different restriction Endo. They cut DNA into fragments of different sizes and numbers. The restriction endonuclease with longer recognition sequence cuts DNA less frequently and produces large fragments in comparisons to enzymes with short recognition sites.

The enzymes Eco R-I and Hind-II have 5 & 7 recognition sequences on DNA. On digestion six and eight fragments of different sizes are obtained respectively. The different restriction enzyme patterns can be observed as distinct bands through agarose gel electrophoresis. Using a mol. wt. marker various fragment size can also be assessed.

PROCEDURE :

   (i)   Place the vials containing restriction enzymes viz. Eco RI and Hind-III on ice.

  (ii)   Thaw the vials containing substrate (Lambda DNA) and assay buffer.

 (iii)   Add 20 µl of λ DNA (substrate) to each of the enzyme vials.

  (iv)   Add 25 µl of 2X assay buffer to the enzyme and DNA mixture, mix by tapping the tube.

   (v)   Incubate the vial at 37°C for 1 hour (or at optimal temperature of the enzyme activity).

  (vi)   Meanwhile, prepare a 1% agarose gel for electrophoresis.

 (vii)   After an hour add 5 µl of gel loading buffer to each of the enzyme vials.

43

(*viii*) Label a 1.5 ml vial as S, add 10 µl of substrate and 1 µl of gel loading buffer to it.

(*ix*) Label a 1.5ml vial as M add 10 µl of marker and 1 µl of gel loading buffer to it.

(*x*) Load the digested samples, substrate and marker, note down the order of loading.

(*xi*) Electrophorese the sample at 50-100 for 2-3 hours.

(*xii*) Stain the agarose gel with IX staining dye.

(*xiii*) Destain to visualize the DNA bands.

OBSERVATION : Observe the band patterns obtained on digestion with Eco RI and Hind-III compare these with molecular weight marker (λ/MI/µl).

A proper digestion should result in sharp bands.

INTERPRETATION : Restriction patterns obtained an digestion with Eco RI and Hind-III are markedly different demonstrating the fact that each restriction enzyme recognizes and cuts only a particular base sequence unique to it.

By comparing the migration distances with that of the marker, one can also determine the approximate size of DNA fragments.

PRECAUTIONS :

(*i*) Enzymes are temp. sensitive, hence place the vials containing enzyme on ice.

(*ii*) Ensure through mixing by gently tapping the vial, after addition of buffer and substrate to the enzyme vial.

(*iii*) Use fresh tip for each addition.

(*iv*) Set the dry bath at 37°C prior to starting the experiment.

(*v*) Care should be taken for preparing 1% agarose gel and 1XTAE buffer.

Restriction digestion
Bgl I (enzyme)

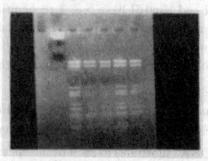

Experiment 14

OBJECT : To perform transformation of the given bacterial population.

MATERIAL REQUIRED –

Host (DN5a) strain, IPTG solution (Isopropyl-B-D Trigadactopyanoside) Xgel, (5 bromo-4 chloro, 3 indoly B-D-Galactopyranoside), plasmid DNA (PUC 18), ampicillin, 1M $CaCl_2$ (Storage 4°C), L.B-medium (R.T.), Centrifuge, incubator, spectrophotometer, conical flask, petriplates, pipettes, spreader, distilled water, capped centrifuge tubes etc.

PRINCIPLE : Bacterial transformation is the process in which bacteria uptake the naked viral or plasmid DNA from surrounding medium. The purpose of this technique is to amplify the plasmid in order to make large quantities of it. Bacteria naturally take up DNA at a certain stage of growth called competence and this can be artificially induced in the cells by treating them with $CaCl_2$ prior to adding DNA. These competent cells allow DNA to enter through pores or channels in the cell membrane and in the case of plasmids, permit subsequent plasmid replication. The Ca^{2+} destabilizes the cell membrane and a Ca-phosphate DNA complex is formed which adheres to the cell surface and is resistant to DNAses. The bacterial cell membrane is permeable to chloride ions but is non-permeable to calcium ions. As the chloride ions enter the cell, water, molecules accompany the charged particle. This influx of water causes to swell and is necessary for uptake of DNA. The DNA is taken up during a heat shock step.

BLUE WHITE TEST : The lac+ bacteria that results from a - complementation are recognized because they form blue colonies in the presence of the chromogenic subserate 5-bromo-4-chloro 3-indodyl B-D-galactoside (X-gel). However, insertion of a fragment of foreign DNA into the polycloning site of the plasmid almost invariable results in production of an amino-terminal fragment that is not capable of α–complementation. Bacteria carrying recombinant plasmids therefore form white colonies.

PROCEDURE :

Day-I : Preparation of Media and Revival of Host –

(*i*) **Preparation of L.B Agar/broth (1 Litre)** - Dissolve 25 gm. of media in 800ml of distilled water. Adjust the pH to 7.0 with 5N NaOH (if necessary) and make up the volume to 1000ml sterilize by autoclaving. For L.B. agar add 1.5% agar and autoclave.

(ii) **Ampicillin Preparation** : Dissolve 100mg of ampicillin in 1ml sterile water to get a stock concentration of 100mg/ml, store at 4°C for 2 weeks.

(iii) **For Ampicillin L.B Media** : Add ampicillin to L.B broth or agar at a final concentration of 100 µg/ml when the temperature of the media is around 40-45°C.

(iv) **Preparation of L.B Ampicillin Plates with X-Gal and IPTG** : After ampicillin is added to the media, add 40 µl each of X-Gal and IPTG for every 20ml of L.B Agar. Mix well and pour media into required number of plates.

(v) **Revival of Host**

(a) Break open the lyophilized vial, add 0.1ml of L.B media.

(b) Streak a loopful of suspenson onto L.B plates.

(c) Incubate the plate at 37°C for 24 hrs.

Day-II :

(i) Inoculate a single colony into 5ml of L-B medium and incubate at 37°C for 24 hrs.

Day-III : Preparation of Competent Cells

(i) Inoculate 1ml of overnight culture into 100ml L.B medium (in a 1 litre conical flask) and incubate at 37°C in a shaker for 2-3 hrs. and check the O.D. (OD reaches 0.23 – 0.26 at A_{600}).

(ii) Chill the culture flask on ice for 10-20 minutes.

(iii) Transfer the culture aseptically into sterile centrifuge tubes and spin at 6000 rpm for 8 minutes.

(iv) Discard the supernatant and to the cell pellet add approx – 15ml of cold 0.1M $CaCl_2$ solution aseptically. Suspend the cell pellet gently in the solution. The tubes are not removed from ice during resuspension.

(v) Place the tube on ice for 30 minutes

(vi) Centrifuge at 6000 rpm for 8 minutes either at 4°C or RT.

(vii) Discard the supernatant and resuspend gently in 0.6ml of cold 0.1M $CaCl_2$ solution.

(viii) Aseptically transfer 100 ml of the above competent cells into 6 prechilled vials. Do all transfers on ice.

(ix) Competent cells are now ready and should be used immediately for the transformation.

Transformation Procedure :

(i) Add 5µl (100ng) of the plasmid DNA to 5 aliquots of 100 µl of competent cells. Gently tap and incubate on ice for 20 minutes (for the DNA to bind to cells). The remaining one aliquot will not be transformed.

(ii) Incubate the vials containing cells in 42°C water bath for 2 minutes, then return vials to ice to chill for 5 minutes.

(iii) Add 1ml of L.B. broth aseptically to the vials and incubate at 37°C for an hour. This is to allow bacteria to recover and express the antibiotic resistance.

(iv) Label three L.B. ampicillin plates with X-Gal and IPTG as a, b and c. Pipette 100 µl of L.B broth on to each plate. Add 25, 50 and 100 µl of transformed cells to plates a, b and c respectively. Mix well and spread thoroughly using a spreader.

(v) Label control plate / non transformed plate. To this plate 100ml of competent cells that has not been transformed, to check for any cell contamination.

(vi) Incubate the plates at 37°C for 24 hrs.

OBSERVATION :

(i) Plate where no DNA added shows no colony.

(ii) Plate with DNA shows colony growth.

CALCULATION :

Transformation efficiency is expressed as No. of transformants / μg of DNA.

$$\text{Transformation Efficieny} = \frac{\text{No. of Colonies x 1000ng}}{\text{Amount of DNA plated (in ng)}} = \underline{\hspace{1cm}}/\mu g$$

e.g. – Amount of DNA transformed = 100ng

Volume of culture plate = 25 μl (of Iml)

Thus, amount of DNA plated = 2.5ng.

If number of colonies observed on plating 2.5ng = 250.

$$\text{Transformation Efficiency} = \frac{250 \times 1000}{2.5}$$

$$= 1 \times 10^5/\mu g.$$

PRECAUTIONS :

(i) The experiment should be done strictly under a septic conditions.

(ii) Revive the strain as soon as the lyophilized vial is opened.

(iii) Always use sterile distilled water.

(iv) Prepare competent cells within 3 days of reviving the strain.

(v) Carry out transformation as soon as the competent cells are prepared.

(vi) Pre-cool tubes, pipettes, 0.1M $CaCl_2$ solution and centrifuge tube / vials, prior to preparation of competent cells.

(vii) Always place the vials / centrifuge tubes containing calcium chloride on ice during resuspension.

(viii) Always cover X-Gal with aluminium foil, as it is light sensitive.

OBSERVATION : Record your observations as follows :

S. No.		Growth	No. of Colonies	Transformation efficiency
1.	Control Plate or Non-transformed Plate			
2.	Transfermed Plate (a) Transfermed Plate (b) Transfermed Plate (c)			

Denote :   + ve : When you observe bacterial growth.

                      – ve : When there is no growth.

INTERPRETATION :

(*i*) On transforming competent cells with pUC 18, antibiotic resistance is conferred on the host as this plasmid carries the gene for ampicillin resistance. As a result, only transformed cells grow on L.B. ampicillin unlike non transformed cells.

(*ii*) These transformed colonies appear blue on L.B ampicillin X-Gal-IPTG plate due to a-complementation i.e. active B-galactosidase produced cleaves X-gal to give the blue colour on IPTG induction.

(*iii*) Using calcium chloride method for preparation of competent cells, the expected transformation efficiency on transforming 100ng of pUC 18 is approximately 1×10^5/ug of DNA. Efficiency lower than this may be attributed to improper condition during preparation of competent cells. *e.g.* Temperature higher than 4°C.

Experiment 15

OBJECT : DNA amplification by Polymerase Chain Reaction.

REQUIREMENTS : Taq DNA polymerase, Deoxynuleotide triphosphate, Assay buffer (100µl 10X Taq Polymerase assay buffer with magnesium chloride composition. 100mM Tris HCl (pH 9.0), 50mM KCl, 15mM $MgCl_2$ and 0.1% gelatin DNA template, Primers (Forward and Reverse Primers), Nuclease free water, Mineral Oil, Agarose, Gel loading dye, PCR Tubes-12, Control DNA Marker-5-ug, 50 X TAE-40ml.

PRINCIPLE : PCR is a method for invitro enzymatic amplification of specific nucleic acids using multiple cycles of template denaturation, primer annealing and primer elongation. The technique is based on discovery of special DNA polymerase enzyme (Taq. polymerase) which are thermostable. These enzymes can tolerate high temperature applied for denaturation of of DNA which generate template DNA.

DNA fragment thus, a 20 cycles of PCR yield about million copies of the target DNA – The doubling of the number of DNA strands corresponding to the target sequences allows us to estimate the amplification associated with each cycle using the formula.

Amplification = 2n

Where, n = Number of Cycles

METHOD :

(*i*) For the reaction add the following reagents to a PCR tube.

    (*a*) 10X Taq. Pol. Assay buffer 15mM $MgCl_2$ – 5µl

    (*b*) dNTP Mix – 3µl

    (*c*) Template DNA – 0.1µl

    (*d*) Forward Primer (250ng/µl) – 0.1µl

    (*e*) Reverse Primer (250 ng/µl) – 1µl

    (*f*) Taq DNA Polymerase (3u/µl) – 1µl

    (*g*) Sterile water – 38 µl

        Total Reaction Volume – 50µl

(*ii*) Mix the contents gently

(*iii*) Carry out the amplification using following the reaction condition for 30 cycles.

		Temp.	Time
(*a*)	Initial Denaturation	94°C	1 Min.
(*b*)	Denaturation	94°C	30 Secs.
(*c*)	Annealing	48°C	30 Secs.
(*d*)	Extension	72°C	1 Min.
(*e*)	Final Extension	72°C	2 Mins.

(*iv*) After the reaction is over, take out the reaction mix. and run 10 µl of it is 1% agarose gel for 2 – 3 hours at 100 volts. Run the sample along with marker and locate the amplified product by comparing with the 0.8 Kb fragment of the marker.

RESULT : Amplified DNA product is obtained.

PRECAUTIONS :

(*i*) Carefully add the required amount of reagent into a PCR tube.

(*ii*) Carry out the amplification using proper reaction condition (Time and Temperature)

(*iii*) Preincubate at 95°C for 5 mins. in the absence of enzyme to inactivate harmful nuclease or protease in the sample.

Experiment 16

OBJECTS : Preparation of Plant Tissue Culture Media.

CULTURE MEDIUM

A substance used to provide nutrients for growth and multiplication of tissue or micro-organisms.

Characteristics of An Ideal Culture Medium : The characteristics of an ideal culture medium are --

 (i) It must give a satisfactory growth from a small inoculum or tissues and ideally from a single cell.

 (ii) It should give a rapid growth.

 (iii) It should be easy to preparae.

 (iv) It should be reasonably cheap.

 (v) It should be easily reproducible.

 (vi) It should make it possible for all the characteristics in which we are interested to be demonstrated.

Ingredients of An Ideal Tissue Culture Medium : An ideal nutrient medium for plant tissue culture contains five classes of ingredients.

 (i) Inorganic salt.

 (ii) Vitamins

 (iii) Carbon source

 (iv) Growth regulators

 (v) Organic supplements

Role of Elements in Nutrient Medium for Tissue Culture Medium :

ELEMENT	FUNCTIONS
Nitrogen	Component of proteins, nucleic acid and some coenzyme.
Potassium	Regulates osmotic potential and principle inorganic cation.
Calcium	Cell wall synthesis, cell signaling and membrane function.
Magnesium	Enzyme cofactor.

51

Phosphorus	Component of nucleic acids, energy, transfer component of intermediates in respiration and photosynthesis.
Chlorine	Required for photosynthesis.
Sulphur	Components of amino acid viz. methionine, cysteine and some cofactors.
Iron	Electron transfer as a component of cytochrome.
Manganese	Enzyme cofactor.
Copper	Enzyme cofactor.
Zinc	Enzyme cofactor, chlorophyll biosenthesis.
Cobalt	Component of some vitamins.
Molybdenum	Component of nitrate reductase and enzyme cofactor.
Sucrose	Carbon source.
Vitamin Thiamine	Improve the cell growth.

GROWTH REGULATORS

Class	Name	Functions
Auxin	Indole-3 Acetic Acid (IAA)	Use for callus induction at 10-30 µm. Lowering to 1-10µm can stimulate organogenesis.
	Indole-3 Butyric Acid (IBA)	Use for rooting shoots, regenerated via organogenesis. Either maintain at low conc. (1-50µM) throughout rooting process or expose to high conc. (100-250µM) for 2-10 days. Can also use as a dip for in vitro or ex-vitro rooting of shoots.
Cytokinin	6-Furfurylamino-purine (Kinetin) K	Promote cell division, callus induction, growth of callus and cell suspension and induction of morphogenesis (1-20µM) higher conc. (20-50µM) can be used to induce the rapid multiplication of shoots, axillary buds or meristems.
	6-Benzylamino-purine (BAP)	Used for callus induction and growth of callus and cell suspensions (0.5–5.0µM) and for induction of morphogenesis (1-10µM)
Gibberellin	Gibberellin A_3 GA_3	Involved in regulating cell elongation, in determining plant height and fruit set.Can promote shoot induction at 0.03-14µM.

Abscisic Acid	Abscisic Acid (ABA)	Inhibits cell division. Used at conc. of 0.04-10μM to prevent precocious germination and promote normal development of somatic emboyos.
Ethylene	Ethylene	Associated with controlling fruit ripening in climacteric fruits and its use in plant tissue culture is not widespread.
Chlorocholine chloride	CCC	Inhibitor of GA_3 biosynthesis, (less commonly used).

PREPARATION OF PLANT CELL CULTURE MEDIA : The main objective of medium preparation is to culture the cell, tissue or organ in vitro.

PROTOCOL

(*i*) Prepare stock solution of major ingredients viz, inorganic salt, vitamins, plant growth regulator and organic nutrients using analytical reagent grade chemicals and sterile distilled water or Packaged powders of the MS salts and other media are available, eliminating the need to prepare stocks and measure ingredients. Follow the manufacturers directions for their use.

(*ii*) To make 1 litre of liquid medium pipette the required volumes of each stock solution into a 1.5 litre beaker on a magnetic stirrer.

(*iii*) Sucrose is added as solid. Stir till it is fully dissolved.

(*iv*) Adjust the volume to 950ml with sterile distilled water.

(*v*) Adjust the pH to the required value (generally pH 5.5 – 6) with 0.5 M NaOH.

(*vi*) Then, transfer to a 1 litre measuring cylinder or volumetric flask and adjust volume to 1 litre with sterile disttled water.

(*vii*) Again transfer back to the flask and stir for complete mixing.

(*viii*) Transfer 75ml batches of medium to 250ml conical flask, plug with cotton, cover the tops with aluminium foil and autoclave 15 min. at 120°C.

PRECAUTIONS

(*i*) Store inorganic stock solution at 4°C for maximum of one month.

(*ii*) Dispense vitamin in 10ml aliquots and it is stored at – 20°C.

(*iii*) Always prepare fresh stock of growth regulator each time.

(*iv*) Auxins are generally titrated into solution with NaOH whereas cytokinins are dissolved in dilute NaOH or aqueous ethanol. ABA stock solution is prepared in water.

(*v*) The pH of plant tissue culture media is generally adjusted to pH 5.5 to 6. Always adjust the pH before adding the agar.

(*vi*) Because of excessive containmination problems with certain plant explants, antibiotics are used in the culture medium viz. Timenlin, Carbenicillin (500mg/litre), Cefotaxime (300μg/ml) and Augmentin (250mg/liter). The antibiotics are soluble in water, should be made up fresh and added to the medium after autoclaving by fitter sterilization.

(*vii*) Any stock that appears cloudy or has precipitates in the bottom should be discarded.

(*viii*) Reagent – grade chemicals should always be used to ensure maximum purity.

PREPARATION OF EXPLANT AND ITS STERILIZATION

Explant is a piece of plant tissue taken from its original site and transferred to an artificial medium (culture medium) for growth. A wide range of plant organs and tissues can be used as a source of explants for the initiation of callus culture. The choice of explant, is generally dictated by the aim of the research viz, if clonal propagation is the aim, then the explant will usually be a lateral or terminal bud or shoot. For callus induction, pieces of the cotyledon, hypocotyle, leaf, stem are generally used.

(*i*) Explant is usually younger tissue.

(*ii*) It is the smaller size (depending on the aim of experiment).

(*iii*) It is advisable to obtain explants from plants which are healthy as compared to plants under nutritional or water stress or plants which are exhibiting disease symptoms.

SURFACE STERILIZATION OF PLANT MATERIAL

It is necessary to effect surface sterilization of the organ from which the tissue is to be aseptically excised or of the spare or seeds whose germination shall yield the tissue explant. Plant material which is to be cultured, should be surface sterilized to remove the surface borne micro-organisms. This procedure is done in front of a Laminar air flow before the plant material is inoculated on to the culture medium. A plant material can be sterilized by any one of the following sterilizing agent.

(*i*) 1-4% solution of sodium hypochlorite (commercial bleach)

(*ii*) 7% saturated solution of calcium hypochlorite

(*iii*) 1% solution of bromine water

(*iv*) 70% ethyl alcohol

(*v*) 0.1 – 0.2% mercuric chloride

(*vi*) 10% hydrogen peroxide solution.

(*vii*) 1% silver nitrate solution

The type, concentration and duration of exposure of the particular sterilant to be used are dependent on the plant material under use.

In many cases, a drop of liquid detergent or wetting agent (Teepol, Tween-80) is incorporated into sterilizing solution to enhance the penetration and effectiveness of the sterilizing agent. Material should be thoroughly washed several times with sterilized distilled water to ensure proper removal of sterilant.

PRECAUTIONS

(*i*) The whole procedure is done in front of laminar air flow

(*ii*) All the sterilized materials and plant parts should not be touched directly by hand and use always sterilized forceps.

(*iii*) Hands and working platform are made aseptic by ethanol (80%) before start the work.

Experiment 17

OBJECT : To isolate protoplast from the plant tissues.

REQUIREMENTS : Young leaves of *Nicotiana tobacum*, 70% ethanol, 0.4 - 0.5% sodium hypochlorite solution, sterile distilled water, sterilized forceps, petri plates, Cell Protoplast Washing Media (CPW), enzyme 2% cellulase (Onozuka R10), 0-5% macerozyme, manitol, Sucrose pasteur pipette, screw capped, centrifuge tube, 60μ - 80μ nylon mesh. Microscope and Laminar air flow.

PRINCIPLE : Isolated protoplast is naked plant cell (without cell wall) surrounded by plasmamembrane, potentially capable of cell wall regeneration, cell division, growth and plant regeneration in culture.

The protoplast can be isolated in viable state by mechanical removal or enzymatic degradation of cell wall. In enzymatic method cells are treated with enzymes cellulase and mcerozyme which degrate the principal constituents of cell wall like pectin, cellulose, hemicellulose etc. In order to avoid injury to protoplasts, cells are first plasmolysed by putting them in a hypertonic solution. Since isolated protoplast is an osmotically fragile system in nascent stage in order to avoid injury due to endosmosis it should be kept in isotonic solution just after its isolation, it also prevents spontaneous regeneration of cell wall by viable protoplast.

METHOD

(*i*) Select young fully expanded leaves from 7-8 week old plants growing in a green house and are washed thoroughly with tap water.

(*ii*) Surface sterilization of leaves is done by first immersing in 70% ethanol for 60 sec. followed by dipping into 0.4 - 0.5% sodium hypochlorite solution for 30 min. For this purpose a sterile casserole dish is used as a sterilizing container, sterilization is done in front of Laminar air flow.

(*iii*) Wash the leaves thoroughly with sterile distilled water to remove every trace of sodium hypochlorite.

(*iv*) Peel the lower epidermis with sterilized fine forceps and cut out the peeled areas with fine scalpel.

OR

Where peeling of the leaf is not possible, slicing of the leaf into thin strips may be sufficient to allow entry of the enzyme through the cut edges of the strip.

(v) Peeled leaf pieces are placed lower surface down onto 30ml sterilize CPW 13M solution in a 14 cm. petridish, Peeled surface is in contact with the solution.

(vi) After about 30 min. replace the mannitol / CPW solution by filter sterilized enzyme solution containing 4% cellulase (onozuka R10), 0.4% macerozyme in 13% manitol added CPW (pH 5-5).

(vii) Leaf pieces in enzyme solution are incubated in the dark at 24 - 26°C for 16-18 hrs.

(viii) Gently squeeze the leaf piece with pasteur pipette to liberate the protoplasts.

(ix) Remove the large debris by filtering through a 60-80 μm nylon mesh.

(x) Transfer the filterate to a screwcapped centrifuge tube and spin at 100 X gm for 3 min.

(xi) The protoplasts form the pellet. Remove the supernatant on the top and the pellet is resuspended in CPW 21S solution it is again centrifuged for 5-10 min. (spin it at 100Xg for 10 min.)

(xii) Collect the green protoplast band from the top of the sucrose pad and transfer it to another centrifuge tube.

(xiii) The viable protoplasts are collected from the surface and are resuspended again in CPW 13M to remove the sucrose. Centrifugation are repeated two-three times for washing.

(xiv) Finally the protoplasts are suspended in measured volume of liquid Nagata and Tabeke medium supplemented with NAA (3mg/L), 6 Benzylamino purine (6-BAP) (1mg/L) and 13% mannitol.

Plate the protoplasts as a thin layer in petri-plates.

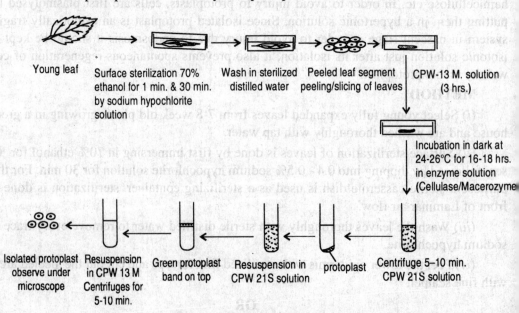

Fig. 12. Protoplast Isolation.

OBSERVATIONS : A clean layer (green in colour) of protoplasts will be seen. Unde low powder objective isolated protoplasts are seen.

PRECAUTIONS :

(*i*) Surface sterilization of leaves should be done properly.

(*ii*) Leaf pieces in enzyme solution are placed for desired period.

(*iii*) Care should be taken while collecting viable protoplast.

Flowchart for Isolation of Protoplasts :

Young Leaves of Healthy Plants

↓

Surface Sterilization of Leaves

(70% ethanol for 1 min and 30 min. by Sodium hypochlorite solution)

↓

Water wash with sterile D/W

(Surface sterilized leaves)

↓

Peeling / Slicing of Leaves

↓

Add Sterilize CPW 13M solution and allowed

it to react for 3 hrs.

↓

Remove the CPW 13M solution by enzyme solution

(Enzyme prepration - cellulase, Macerozyme etc.)

↓

Peeded Leaves + Enzyme Solution

(Incubated in dark at 24-26°C for 16-18 hrs.)

↓

Gently squeeze the leaf pices to Librate protoplast

↓

Remove Large debris. Spin the filterate at

100Xgm for 3 min.

↓

Pellet + Supernalant

(Protoplast) (Remove)

↓

Resuspended in CPW 21S solution and

centrifuge for 5-10 min.

↓

Green Protoplast band on top

↓

Again resuspended in CPW 13M and Centrifuge

↓

Isolated Protoplast

OBJECT : Viability testing of protoplast.

MATERIAL REQUIRED : Fluorescein diacetate, acetone, isolated protoplast, test tubes, droper, centrifuge tube.

PRINCIPLE : In protoplast technology only viable protoplast are subjected to genetical manipulation or plant regeneration, therefore test of viability of isolated protoplast is an important step. In FDA method live cells cleave FDA by esterase activity and produce fluorescein, which cannot cross plasma membrane.

METHOD :

(*i*) Prepare FDA solution by dissolving in 5.0mg/ml acetone and stored at 0°C.

(*ii*) Take the isolated protoplast in centrifuge tube add 1ml of FDA solution into it.

(*iii*) Shake well. The viable protoplast shows fluoresces red colour while dead protoplast remains as such.

(*iv*) Calculate the percentage of viable protoplast in a preparation.

OBSERVATIONS :

S. No.	No. of Protoplast	No. of Viable Protoplast	No. of Dead Protoplast	% of Viability

Viable protoplast shows fluoresces red in colour.

PRECAUTIONS :

(*i*) FDA solution should be prepared carefully.

(*ii*) Within 10-20 minutes isolated protoplast must be checked for its viability.

Experiment 19

OBJECT : Demonstration of protoplast fusion using polyethylene glycol.

MATERIAL REQUIRED : Polythylene glycol, isolated protoplast of different varieties, 1.5ml vials, microscope, slides, droper, pasteur pipette.

PRINCIPLE : Protoplast fusion is a physical phenomenon. During fusion two or more protoplasts come in contact and adhere with one another in presence of fusion inducing chemical PEG. After adhesion, membranes of protoplasts fuse in some localized areas and eventually the cytoplasm of the two protoplasts intermingle. The adhesion of isolated protoplast takes place either due to reduction of negative charges of protoplast or due to attraction of protoplast by electrostatic forces caused by chemical fusogens.

METHOD :

(*i*) Mix freshly isolated protoplast of the two desired parents in a ratio of 1:1 into a 1.5ml vial.

(*ii*) Add drop by drop 450µl of PEG solution to the protoplast suspension.

(*iii*) Incubate the protoplast in PEG solution for 15-30 min. at room temperature (24°C).

(*iv*) With the help of pasteur pipette place the drop of protoplast on to the slide and observe under microscope.

OBSERVATION : The membranes of protoplasts are fused in some localized areas and the cytoplasm of the two protoplasts intermingle.

PRECAUTIONS :

(*i*) Freshly isolated protoplast are used.

(*ii*) Isolated protoplast in the ratio 1:1 is used for the experiment.

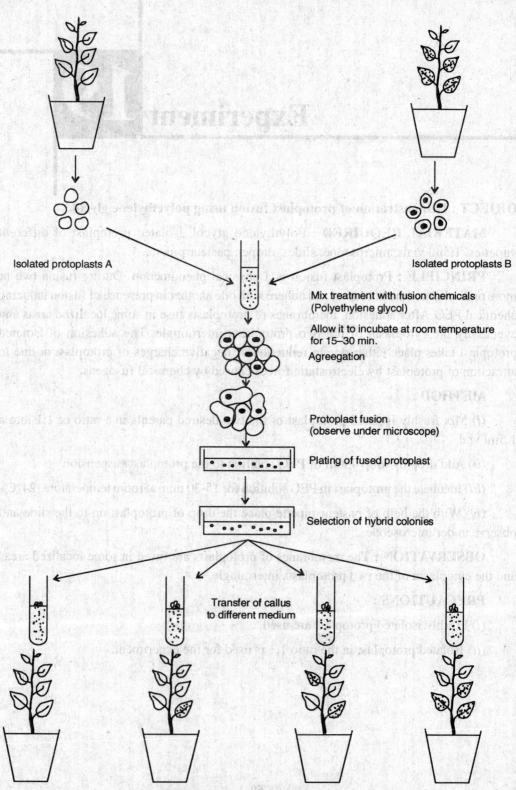

Isolated protoplasts A

Isolated protoplasts B

Mix treatment with fusion chemicals
(Polyethylene glycol)

Allow it to incubate at room temperature
for 15–30 min.

Agreegation

Protoplast fusion
(observe under microscope)

Plating of fused protoplast

Selection of hybrid colonies

Transfer of callus
to different medium

Fig 13. Protoplast Fusion

FLOWCHART FOR PROTOPLAST FUSION

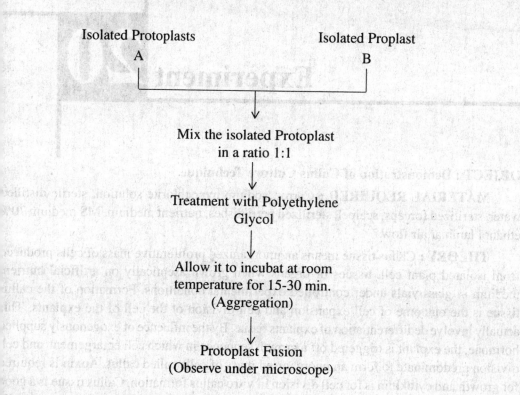

Isolated Protoplasts
A

Isolated Proplast
B

Mix the isolated Protoplast
in a ratio 1:1

Treatment with Polyethylene
Glycol

Allow it to incubat at room
temperature for 15-30 min.
(Aggregation)

Protoplast Fusion
(Observe under microscope)

Experiment 20

OBJECT : Demonstration of Callus Culture Technique.

MATERIAL REQUIRED : Carrot, sodium hypochlorite solution, sterile distilled water, sterilized forceps, scalpel, sterilized petri dishes, nutrient medium, MS medium 70% ethanol laminar air flow.

THEORY : Callus tissue means an unorganized proliferative mass of cells produced from isolated plant cell, tissues or organs when grown aseptically on artificial nutrient medium in glass vials under controlled experimental conditions. Formation of the callus tissue is the outcome of cell expansion and cell division of the cell of the explants. This actually involve dedifferentiation of explants tissue. By the influence of exogenously supplied hormone, the explant is triggered off a growth sequence in which cell entargement and cell division predominate to form an unorganised mass of cells called callus. Auxin is required for growth and cytokinin is for cell division in vitro callus formation. Callus tissue is a good source of genetic variability, so it may be possible to regenerate a plant from genetically variable cells. Callus culture is very useful to obtain commercially important secondary metabolities.

METHOD :

(i) A fresh carrot is taken and washed throughly under running tap water to remove all surface detritus.

(ii) Then, it is dipped into 5% Teepol for ten minutes and then the root is washed.

(iii) The tap root is surface sterilized by immersing in 70% ethanol for 60 sec. followed by 30 min. in sod. hypochlorite (0.8% available chlorine).

(iv) Then, the root is washed three times with sterile distilled water to remove completely the sod. hypochlorite.

(v) The carrot is then transferred to a sterilized petridish. A series of transverse slice 1mm in thickness is cut from the tap root using a sharp scalpel.

(vi) Each piece is transferred to another sterilize petridish. Each piece contains a whitish circular ring of cambium around the pith. An area of 4mm² across the combium is cut from each piece so that each small piece contains part of the phloem, cambium and xylem. Size and thickness of the explants should be uniform.

(*vii*) Always the lid of petridish is replaced after each manipulation.

(*viii*) An explant is transferred using forceps onto the surface of the agarified nutrient medium OR MS medium OR in culture to be containing nutrient media.

(*xi*) Cultures are incubated in dark at 25°C.

(*x*) Usually after 4 weeks in culture the explants incubated on medium with 2,4-D will form a substantial callus. The whole callus mass is taken out aseptically on a sterile petridish and should be divided into 2 or 3 piece.

(*xi*) Each piece of callus tissue is transferred to a tube containing fresh some medium.

(*xii*) Prolonged culture of carrot tissue produces large callus.

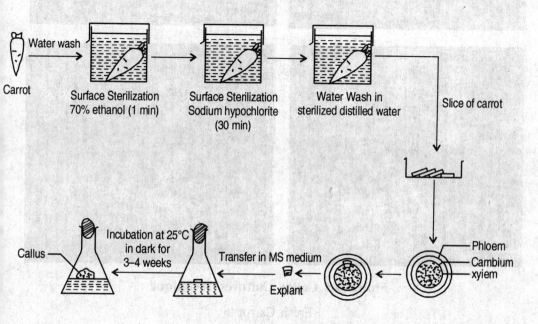

Fig. 14 Callus Culture

OBSERVATIONS : After 4 - 6 weeks callus is farmed. Note the colour, texture, xternal morphology.

(*i*) Colour......................

(*ii*) Texture.........................

(*iii*) External morphology.................

PRECAUTIONS :

(*i*) The whole experiment is performed under aseptic condition.

(*ii*) The forceps, scalpels must be kept in 95% ethanol and flamed throughly before se.

(*iii*) Size and thickness of the explants should be uniform.

(*vi*) Each explant always have a part of phloem, cambium and xylem.

(*v*) Always used sterilized glass wares.

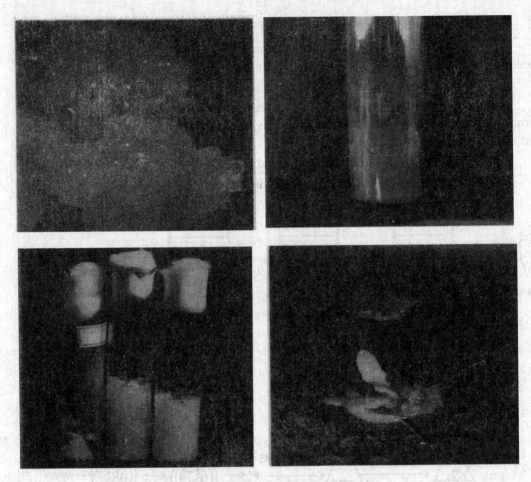

Flowchart Callus Cultures Technique :

Fresh Carrot

Water wash the Fresh Carrot

Surface
Sterilization

Surface sterilization (70% ethanol for 1min and
30 min by sodium hypochlorite solution)

Water Wash with sterile D/W
(Surface Sterilized Carrot)

Transfer to Sterilized Petridish and cut
the transverse slice 1mm thickness

↓

Preparation of Explant — Transfer it to another petridish and than cut an aera
of 4mm² across the cambium from each piece
(Each small piece contain a part of phloem,
cambium and xylem)

↓

Explant is transferred on to MS medium

↓

Incubate it in dark at 25°C for 4 weeks

↓

Callus formation is seen

Experiment 21

OBJECT : Initiation of Callus culture from radical tip of *Phaseolus vulgaris* (French Bean).

MATERIAL REQUIRED : Laminar air flow, *Phaseolus vulgaris*, sodium hypochlorite solution, 70% ethanol, sterile distilled water, sterilize forceps, scalped, petridishes, M.S. medium etc.

THEORY : Callus tissue means an unorganized proliferative mass of cells produced from isolated plant cell, tissues or organs when grown aseptically on artificial nutrient medium in glass vials under controlled experimental conditions. Formation of the callus tissue is the outcome of cell expansion and cell division of the cell of the explants. This actually involve dedifferentiation of explants tissue. By the influence of exogenously supplied hormone, the explant is triggered off a growth sequence in which cell entargement and cell division predominate to form an unorganized mass of cells called callus. Auxin is required for growth and cytokine is for cell division in vitro callus formation. Callus tissue is a good source of genetic variability, so it may be possible to regenerate a plant from genetically variable cells. Callus culture is very useful to obtain commercially important secondary metabolites.

METHOD :

(*i*) Place seeds (15 seeds/petridish) in petridishes.

(*ii*) Fill petridishes with 70% ethanol for 1 min followed by 30 min in sodium hypochlorite.

(*iii*) Wash the seeds in the petridish with 3 - 4 times by sterile distilled water.

(*iv*) Leave seeds in petridish to seak overnight in sterile distilled water.

(*v*) By using sterile forceps discard any seeds with cracked testas.

(*vi*) Re-sterilize the seeds in sodium hypochlorite for 20 min.

(*vii*) Again wash the seeds with 3 - 4 times by sterile distilled water.

(*viii*) Transfer the seeds (5 seeds / petridish) to sterile petridishes.

(*ix*) By using forceps to hold the seed, then make two cuts with scalpel in order to allow the testa to be pulled to reveal the radical.

(*x*) Cut out the radical tip (2-3 min) and transfer to another petridishes containing sterile distilled water (15 tips / dish).

(*xi*) Then transfer the radical tips to MS medium (5 tips / petridish).

(*xii*) Seal the dishes with parafilm and incubated in dark at 25°C for 4 - 6 weeks.

OBSERVATIONS : Callus is formed. Note the colour, texture and external morphology.

PRECAUTIONS :

(*i*) The whole experiment is performed under aseptic condition.

(*ii*) The forceps, scalpels must be kept in 95% ethanol and flamed throughly before use.

(*iii*) Size and thickness of the explants should be uniform.

(*iv*) Each explant always have a part of phloem, cambium and xylem.

(*v*) Always used sterilized glass wares.

Fig. 15

Experiment 22

OBJECT : Demonstration of cell suspension culture.

MATERIAL REQUIRED : Laminar air flow Callus tissue, conical flasks, MS Medium, centrifuge, test tubes, shaker incubator or rototary shakes, haemocytometer, pipettes, aluminium foil, cotton, brown paper, nylon mesh, sieve (pare diameter 60μ - 100μ) microscope, laminar air flow, pipette.

PRINCIPLE : Cell suspension culture is a type of culture in which single cell or small aggregates of cells multiply while suspended in agitated culture medium. Such technique gives valuable information about cell physiology, biochemistry, metabolic events at the level of individual cells or of small cell aggregates and also important for understanding of an organ formation or embryoid formation. It is used for the whole or partial synthesis of secondary plant products.

METHOD :

(*i*) Take 250ml conical flask containing 100ml of MS Medium.

(*ii*) Transfer 3-4 pieces of pre-established callus tissue (approx. wt. 1gm. each) to conical flask containing media.

(*iii*) Place the flask within the clamps of a shaker at 80-120 rpm. and incubated for 7 days.

(*iv*) After 7 days pour the content of flask to sterilize sieve and collect the filterate in a sterile conical flask. The filtrate contains only free cells and cell aggregates.

(*v*) Allow the filtrate to settle for 15-20 mins OR centrifuge the filtrate to 500-1000 rpm and finally pour of the supernatant.

(*vi*) Resuspend the residue cells in a requisite volume of fresh liquid medium and then dispense the cell suspension equally in 3-4 sterilized flasks.

(*vii*) Place the conical flasks on shaker and allow the free cells and cell aggregate to grow.

(*viii*) Repeat the previous steps for sub-culturing into fresh nutrient medium 3-4 time (liquid) take only 10ml of cell suspension.

(*xi*) Pipette out very little aliquot of cell suspension and count the cell number with the help of haemocytometer under microscope.

(*x*) Note the various shapes and sizes of cells the cell number per unit volume of culture in relation to time (days) of culture.

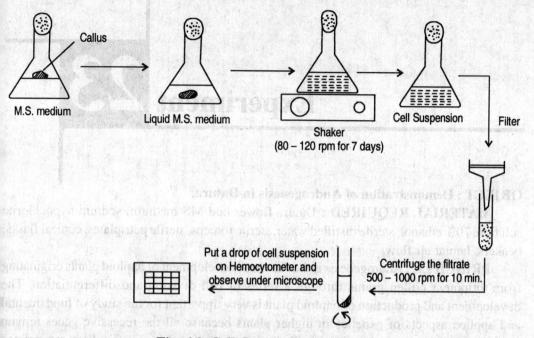

Fig. 16. Cell Suspension Culture.

OBSERVATIONS :

(*i*) Shapes of the Cell are............................

(*ii*) Cell No. is..

PRECAUTIONS :

(*i*) The whole experiment should be perform under aseptic condition.

(*ii*) Always used sterilized glassware.

OBJECT : Demonstration of Androgenesis in Datura.

MATERIAL REQUIRED : Datura flower bud MS medium, sodium hypochlorite solution, 70% ethanol, sterile distilled water, sterile forceps, sterile petriplates, conical flasks, beakers, lamiar air flow.

PRINCIPLE : Androgenesis is the "in vitro" development of haploid plants originating from totipotent pollen grains through a series of cell division and differentiation. The development and production of haploid plant is very important for the study of fundamental and applied aspects of genetics in higher plants because all the recessive genes remain uncovered and the recessive gene controlled traits are expressed phenotypically in regenerated plant. Haploids are very useful for plant breeding programmes.

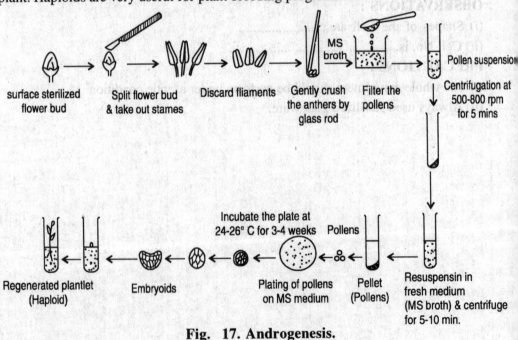

Fig. 17. Androgenesis.

METHOD (a) :

(i) Collect the flower buds (Datura) and washed throughly under running tap water to remove all surface detritus.

(*ii*) Then, flower buds are surface sterilized by immersing in 70% ethanol for 60 sec. followed by 20-30 min. in sodium hypochlorite solution.

(*iii*) Washed it with three to four times sterile distilled water to remove completely the sodium hypochlorite.

(*iv*) Finally, transfer the buds to sterile petridish.

(*v*) To remove another, slit the side of the bud with a sharp scalpel and remove them with a pair of forceps, place the anthers with filaments to another petridish.

(*vi*) The filaments are cut gently. During excision of anthers, special care should be taken to ensure that they are not injured in any way. Damaged anthers should be discarded.

(*vii*) Anthers are placed on agar solidified basal MS medium.

(*viii*) The culture are incubated at 24-28°C in dark for 3-4 weeks.

OR

METHOD (*b*) :

(*i*) Collect the flower buds (Datura) and washed throughly under running tap water to remove all surface detritus.

(*ii*) Then, flower buds are surface sterilized by immersing in 70% ethanol for 60 sec. followed by 20-30 min. in sodium hypochlorite solution.

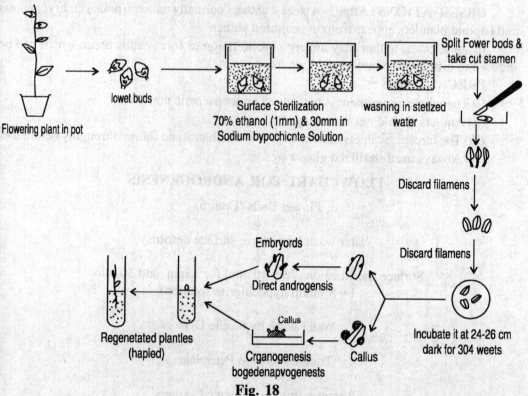

Fig. 18

(*iii*) Washed it with three to four times sterile distilled water to remove completely the sodium hypochlorite.

(*iv*) Finally, transfer the buds to sterile petridish.

(*v*) To remove anthers, slit the side of the bud with a sharp scalpel and remove them with a pair of forceps, place the anthers with filaments to anthers petridish.

(*vi*) The filaments are cut gently. During excision of another, special care should be taken to ensure that they are not injured in any way. Damaged anothers should be discarded.

(*vii*) About 50 anthers are placed in small sterile beaker containing 20ml of liquid basal MS medium.

(*viii*) Anthers are then pressed against the side of the beaker with sterile glass piston of a syringe to squeeze out the pollens.

(*ix*) The homogenized anthers are then filtered through a nylon sieve to remove the another tissue debris.

(*x*) The filtrate or pollen suspension is then centrifuged at low speed (500-800 rpm) for 5 minutes. The supernatant containing fine debris is discarded and the pellet of pollen is suspended in fresh liquid MS medium and washed twice by reputed centrifugation and resuspension in fresh MS liquid medium.

(*xi*) Pollens are mixed finally with measured volume of liquid basal medium.

(*xii*) 2.5ml of pollen suspension is piptted off and is spread in 5 cm. petridish, containing liquid medium OR plating in very soft agar added medium.

(*xiii*) Petridishes are incubated at 27-30°C for 3-4 weeks.

OBSERVATIONS : After 3 - 4 weeks, anthers normally undergo pollen embryogenesis and haploid plantlets appear from the cultured anther.

In some cases, anther may undergo proliferation to form callus tissue which can be induced to differentiate into haploid plants.

PRECAUTIONS :

(*i*) The whole experiment is performed under asepetic condition.

(*ii*) Anthers should not be damaged.

(*iii*) The forceps, scalpels must be kept in 95% ethanol and flamed throughly before use

(*iv*) Always used sterilized glasswares.

FLOWCHART FOR ANDROGENESIS

Flower Buds (Datura)

↓

Water wash (to remove surface detritus)

↓

Surface Sterilization (70% ethanol for 1 min. and 30 min.
by Sodium hypochlorite solution)

↓

Water wash by Sterile D/W

↓

Transfer buds to Petriplate

↓

Remove anthers, Damaged anthers
are discarded

↓

Transfer the anthers onto agar solidified MS
medium (Culture Medium)

$\downarrow$

Incubated at 24 - 28°C in dark for
3 - 4 weeks

$\downarrow$

Haploid Plantlets appear

OR

02. METHOD OF ANDROGENESIS -

Flower Buds (Datura)

$\downarrow$

Water wash (to remove surface detritus)

$\downarrow$

Surface Sterilization (70% ethanol for 1 min. and 30 min.
by Sodium hypochlorite solution)

$\downarrow$

Water wash by Sterile D/W

$\downarrow$

Transfer buds to Petriplate

$\downarrow$

Remove anther, Damaged anthers
are discarded

$\downarrow$

Transfer anthers to MS broth

$\downarrow$

Pressed the anther to squeeze out pollens

$\downarrow$

Filtered it

$\downarrow$

Pollen suspension or filterate centrifuge
at 500 - 800 rpm for 5 mins.

$\downarrow$

Pellet + Supernatant (discarded)
(Pollen suspension)

$\downarrow$

Add MS broth and Centrifuge 5-10 mins.
(Repeat the step of centrifugation twice)

$\downarrow$

Transfer the pallen on to MS medium OR
Plated in soft agar MS medium

$\downarrow$

Incubated the plates at 24 - 32°C for
3 - 4 weeks

$\downarrow$

Haploid Plantlets appear

OBJECT : Demonstration of Somatic embryogenesis.

MATERIAL REQUIRED : For explant - Daucus carota.

CHEMICALS : Murashige and skoog's media (M.S.), M.S. Broth, Sodium Hypochlorite solution, 70% ethanol. Glasswares - Sterile petriplates, conical flask, beakers and laminar air flow.

OTHER REQUIREMENTS : Scalpel, forceps, distilled water.

PRINCIPLE : A somatic embryo (SE) is an embryo derived from a somatic cell, other than zygote usually on culture in vitro and the process is known as somatic embryogenesis. The capability of the somatic plant cell of a culture to produce embryoids is known as embryogenic potential, somatic embryo development needs some prior callus formation and embryoids (a small, well organised structure which is produced in tissue culture of dividing embryogenic potential somatic cells), originate from induced embryogenic cells within the callus tissue.

METHOD :

(*i*) A fresh carrot is taken and washed throughly under running tap water to remove all surface detritus.

(*ii*) Then, it is dipped into 5% Teepol for ten minutes and then again water wash it.

(*iii*) The tap root is surface sterilized by immersing in 70% ethanol for 60 sec. followed by 20 - 30 min. in sodium hypochlorite.

(*iv*) Then, the root is washed three times with sterile distilled water to remove completely the sodium hypochlorite.

(*v*) The carrot is then transferred to a sterilized petridish a series of transverse slice 4mm in thickness is cut from the tap root using a sharp scalpel.

(*vi*) Each piece is transferred to another sterile petridish. Each piece contains a whitish circular ring of cambium around the pith. An area of 4mm² across the cambium is cut from each piece so that each small piece contains part of the phloem, cambium and xylem. Size and thickness of the explants should be uniform.

(*vii*) Always the lid of petriplate is replaced after each manipulation.

(*viii*) Following aseptic technique an explants are transferred individually using forceps on a semi solid Musashige and skoog's media contains 2% sucrose.

(*ix*) Cultures are incubated in dark. In this medium the explant will produce sufficient callus tissues.

(*x*) After 3 - 4 weeks of callus growth, cell suspension culture is to be initiated by transferring 0.2 gm. of callus tissue to a 250ml of Erlenmeyer flask containing 20-25ml of liquid medium of the same composition are used for callus growth (without agar) Flasks are placed on a shaker with 125-160 rpm at 25°C.

(*xi*) Cell suspensions are subcultured every 4 weeks by transferring 5ml to 65ml of fresh MS medium. Cultures are incubated in dark.

(*xii*) After 3 - 4 weeks of culture contain numerous embryos in different stages of development.

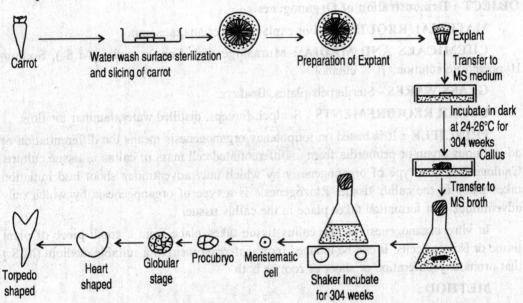

Fig. 19. Somatic Embryogenesis.

OBSERVATIONS : Numerous embryos in different stages of development are seen. They are globular (round ball shaped), heart shaped, torpedo and lastly the cotyledonary stage. The torpedo stage is a bipolar structure which ultimately gives rise to complete plantlet.

PRECAUTIONS :

(*i*) The whole experiment is performed under aseptic condition.

(*ii*) The forceps, scalpels must be kept in 95% ethanol and flamed and throughly before use.

(*iii*) Size and thickness of the explants should be uniform.

(*iv*) Each explant always have a part of phoem, cambium and xylem.

(*v*) Always used sterilized glassware's.

OBJECT : Demonstration of Organogenesis.

MATERIAL REQUIRED : For explant - Nicotiana tabacum.

CHEMICALS AND MEDIA : Murashige and skoog's media (M.S.), Sodium Hypochlorite solution, 70% ethanol.

GLASS WARES - Sterile petriplates, Beakers.

OTHER REQUIREMENTS : Scalpel, forceps, distilled water, laminar air flow.

PRINCIPLE : It is based on totipotency organogenesis means the differentiation of adventitious organ or primordia from undifferentiated cell mass of callus in tissue culture. Caulogenesis is a type of organogenesis by which only adventitious shoot bud initiation takes place in the callus tissue. Rhizogenesis is a type of organogenesis by which only adventitious root formation takes place in the callus tissue.

In vitro organogenesis in the callus tissue takes place from a small piece of plant tissue or isolated cells. It can be induced by transferring them to a suitable medium (M.S.) that promote proliferation of shoot or root or both.

METHOD :

(*i*) The upper part of the stem of tobacco plants are harvested and cut into 2 cm. long segments.

(*ii*) Surface sterilization of the tissue is done by immersing the stem piece in 70% ethanol for 30 sec. followed by a 15 minutes in sodium hypochlorite. Then the step piece is washed several times with distilled water.

(*iii*) The stem explants are taken in a sterilized petri-dish and cut longitudinally into two equal pieces and inoculated on to Murashige and skoog's solid medium.

(*iv*) The cultures are then incubated at 25°C with an illumination of about 2000 lux (16 hrs. photoperiod).

(*v*) Callus tissue which is white / yellow in colour, begin to form in two weeks and after six weeks it should be sub-cultured to fresh medium.

(*vi*) Roots or shoots or both primordia develop within 3 weeks of transfer of callus to MS medium.

OBSERVATION : Root or shoots or both primordia developed.

PRECAUTIONS :

(*i*) The whole experiment is performed under aseptic condition.

(*ii*) The forceps, scalpels must be kept in 95% ethanol and flamed throughly before use.

(*iii*) Size and thickness of explants should be uniform.

(*iv*) Always used sterilized glass wares.

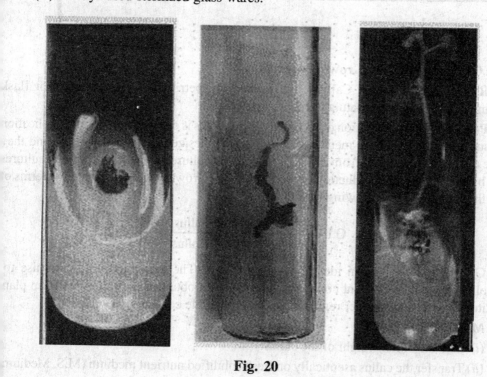

Fig. 20

Experiment 26

OBJECT : Estimation of Growth Index.

    REQUIREMENTS : Callus culture, balance, petriplates, culture tubes or flask containing M.S. medium, spatula, forceps, scalpels.

    PRINCIPLE : Production of secondary metabolites is one of the important application of tissue culture. Secondary metabolites are also synthesized by cells in culture and they can be obtained quite easily from callus or suspension culture. Growth rate of these cultures is the basis of economic production of these products. Growth rate is calculated in terms of Growth Index (G.I.) by following formula –

$$G.I. = \frac{\text{Final Weight of Callus}}{\text{Initial Weight of Callus}}$$

    Growth Index gives an idea about growth rate. The technique holds promise fo controlled production of plant constituents and we can obtained absolutely uniform plan constituents at all times under regulated and reproducible conditions.

    METHOD :

    (*i*) Take the initial weight of the callus.

    (*ii*) Transfer the callus aseptically on fresh solidified nutrient medium (M.S. Medium and incubated in dark at 26-28°C for 2 – 3 weeks.

    (*iii*) Then again take the final weight of callus.

    (*iv*) Calculate the growth rate in terms of Growth Index.

    (*v*) Use these static cultures for detection of plant metabolites and studies in biproductio of secondary products.

    RESULTS : Growth rate is directly proportional to production of secondar metabolites.

    PRECAUTIONS :

    (*i*) Perform all the operations under aseptic conditions.

    (*ii*) Incubate the culture under controlled physical condition.

    (*iii*) Callus can be maintained continuously by serial subcultures.

Experiment 27

AIM : Preparation of artificial seeds by dropping method.

REQUIREMENTS : Callus MS medium, hydrogel sodium alginate, sterite distilled water, petri dishes, laminar air flow.

PRINCIPLES : Artificial seed is a somatic embryo incapsulated within a gel along with nutrients, growth regulators, pesticides, antibiotic etc. It can be used as a substitute of natural seed. Water soluble hydrogel such as sodium alginate, calcium alginate, with gelatin.

METHOD : It involves following steps :

(*i*) Establishment of callus culture.

(*ii*) Induction of somatic embryogenesis in callus culture.

(*iii*) Maturation of somatic embryo.

(*iv*) Encapsulation of somatic embryo.

1. Establishment of callus culture and the induction of somatic embryogenesis in callus culture has already been discussed in experiment number 20 and 24.

2. Maturation of somatic embryo – The embryoids in initial stage are globular, than they become heart shaped and finally they become torpedo – shaped. In final stage embryo attains maturity.

3. Encapsulation of somatic embryo – by Dropping method –

(*a*) Isolated somatic embryos are taken and are mixed with 0.5 to 5% sodium alginate solution and dropped in 30-100 μm calcium nitrate solution, in such a way that each drop contains a singe somatic embryo.

(*b*) Allow it to react for 30 minutes at room temperature. The drop are gelled completely due to formation of calcium alginate.

PRECAUTIONS :

1. Mature somatic embryo should be taken for preparation of artificial seeds.

2. Care should be taken while mixing sodium alginate solution and when dropped in calcium nitrate solution.

FLOW CHART : Dropping Method

Establishment of callus culture

↓

Induction of somatic embryogenesis

↓

Maturation of somatic embryo

↓

Encapsulation of somatic embryos with sodium
alginate and dropped in calcium nitrate

↓

Allow it to react for 30 min.

↓

Artificial seeds are formed

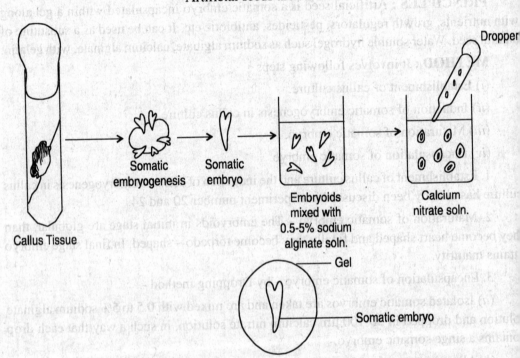

Callus Tissue

Somatic
embryogenesis

Somatic
embryo

Embryoids
mixed with
0.5-5% sodium
alginate soln.

Calcium
nitrate soln.

Dropper

Gel

Somatic embryo

Fig. 21. Synthetic Seed.

Experiment **28**

AIM : Transfer the plantlets to greenhouse condition / field.

REQUIREMENT : Polyethylene tent, humidifiers (e.g. sprinklers, fog unit, Ultrasonic evaporator) special box.

PRINCIPLE : For successful transfer of plantlet to greenhouse condition most of the species require acclimatization in order to ensure their survival ex vitro. The physiological and anatomical characteristics of micro propagated plantlets necessitate that they gradually acclimatize to the environment of the green house or field.

METHOD : Several techniques can be applied for transfer the plantlets to green house/field.

(*i*) The process of acclimatization can start in vitro. Bottom cooling reduces the relative humidity in the head space of the container and this can initiate the weaning process.

(*ii*) Uncap the culture vessels and then put them in green house several days prior to removal of the plants from the culture medium. Contamination of the medium does not become problematic unless plantlets remain in the open vessels for more than one week.

(*iii*) Wash the agar thoroughly from the plantlets because agar serve as a substrate for the growth of disease causing microorganisms.

* When the shoots are rooted in vitro, transplant them carefully to reduce the risk of wounding.

* In case where the roots are newly induced, treat the shoots as cuttings and strike them directly.

* When the roots have not been induced, dipping of the plant base in the rooting powder or solution before striking can be sufficient.

* Unfertilized peat which not wet a good substrate. Rockwool plugs and other inert material are also useful.

(*vi*) Individual plantlets are separated out and transplanted into pot (75 mm) containing seedling compost.

A screen in the green house prevents temperature peaks and lowers the light intensity, which makes the transition easier.

(*v*) After 2 weeks begin acclimating the plants to lower humidity by loosening the plastic around the pot and allowing greater air circulation.

(*vi*) After 8 weeks a full root system should be visible throughout the pot and after that plants are transplanted to soil with a balanced fertilizer. After transplantation, water the plants.

PRECAUTIONS :

(*i*) Care should be taken not to damage root or shoot system when plantlets are removed from the culture vessels.

(*ii*) The plantlets are carefully washed with tap water to remove the residual agar medium.

(*iii*) Maintaining a high relative humidity for the first few days is critical.

(*iv*) The finer the water droplets (10 mm or less) the better, as this avoids a too wet and consequently too anaerobic substrate.

(*v*) The plantlets must initially be kept quite humid to avoid desication.

(*vi*) Potted plants must be protected for about 2 weeks or until the roots are visible in the holes of the bottom of the pots.

Experiment 29

AIM : Isolation of genomic DNA from plant cells.

REQUIREMENTS : Sodium dodesyl sulphate (SDS) EDTA, ysozyme triton X100, MP-14, Liquid nitrogen, Proteinase K (20 mg/ml in water), chloroform, Isoamyl alcohol, DNAse free RNAse, Extraction solution (ES) (NaCl, Tris HCl, Incubator, micro centrifuge, Ice, micropipettes sterile tip, blotting paper, 1.5 ml vials.

1.	Detergents	SDS 10%	Lyse cells and assist in removal of proteins from DNA.
		EDTA	
		Lysozyme	
		Trilon X 100	
		NP – 14	
2.	Liquid N_2		Induce efficient proteolysis and inactivate nucleases.
3.	Proteinase K 20 mg/ml in water		Digest proteins.
4.	Chloroform/isoamyl alcohol (24.1)		Remove proteins, DNA and RNA are not soluble in these solvents.
5.	DNAse free RNAse 1 mg/ml		Remove RNA
6.	Extraction Solution (ES) 0.1 MEDTA, 0.2 MNaCl 0.05 Tris HCl (pH-8) 0.5% SDS, 50 mg/ml RNAse		Dissolve DNA

PRINCIPLE : Cells are first broken and then DNA is separated from other components such as proteins, RNA, liquids and carbohydrates cells are lysed by treating with a detergent SDS. SDS denatures proteins when mixture is denatured the chromosomal DNA is neutralized by addition of sodium acetate, renaturation occurs. DNA remains in solution and can be pepted by using ethoanol/isopropanol.

The detergent SDS carries positive charge which binds with negatively charge DNA n high salt conc. to form soluble complex. Subsequent decreases in salt conc. precipitates DNA leaving other components specially polysaccharides in solution.

METHOD :

(*i*) Plant tissue is grinded in liquid nitrogen with alumina powder.

(*ii*) Take fine powder (approx. 1g. = 1 ml) and to it add 10 volume of Extraction solution.

(*iii*) Incubate the test tubes as 55^0 C for 2-3 hours.

(*iv*) After incubation period add equal volume of phenol/chloroform.

(*v*) Centifuge for 5 minute at 1500 g discard pellet and take supernatent.

(*vi*) To the supernatent add equal volume of chloroform/isoamyl alocohol.

(*vii*) Add 2 volume of ethanol and sodium acetate. Here, DNA is precipitates.

(*viii*) Then, centrifuge (10000 g for 5 min.). Discard supernatent and to the pellet add 70% ethanol.

(*ix*) Dried under vacume desicator/air direct at 37°C.

(*x*) Add tris buffer to get DNA sample.

PRECAUTIONS :

(*i*) Care should be taken while grinding plant tissues in liquid nitrogen.

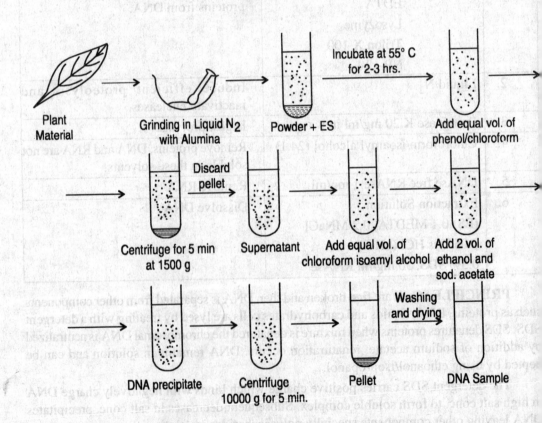

Fig. 22.

Flow – Chart

SDS Proteinase K. method Latest and commonly used method applicable to most materials

Plant tissue

↓

Grinding in liquid N_2 with aluminium powder

↓

Fine powder (1g = 1ml)

↓

Add 10 vol. of ES

↓

Incubate at 55^0 C for 2-3 hrs.

↓

Add equal vol. of phenol / chloroform

↓

Centrifuge for 5 min of 1500 g.

↓ ——————— Discard Pellet

Supernatant

↓

Add equal vol. of chloroform/isoamyl/alcohol

↓

Add 2 vol. of ethanol and sodium acetate

↓

DNA precipitates

↓

Centrifuge (10,000 for 5 min.) Discard Supernatent

↓

Pellet

↓

Washing in 70% ethanol

↓

Dried under vacume deciacator/air dried at 37^0C

↓

Add Tris buffer

↓

DNA sample

AIM : Isolation of *Agrobacterium* sps. **from soil and study of growth characteristics by using plating technique.**

REQUIREMENTS : Agrobacterium agar medium plates, agrobacterium broth, test tube, nicron wire loop.

PRINCIPLE : *Agrobacterium* is a rod shaped gram negative, soil borne, motile bacteria when it grown on a variety of media, will exhibit difference in the macroscopic appearance of their growth. These differences, called cultured characteristics which are used as base for separating Agrobacterium into taxonomic groups.

METHOD : Prepare a *Agrobacterium agar* plates broth and *Agrobacterium tumefaciens* selective medium. Agrobacterium Agar – composition per liter.

Agar	15.0 g
Mannitol	8.0 g
NaCl	5.0 g
Yeast extract	5.0 g
$(NH_4)_2 SO_4$	2.0 g
Casamino acids	0.5 g

pH 6.6 ± 0.2 at 25^0 C

Agrobacterium tumefaciens selective medium composition per 1020 ml

Agar	15.0 g
L (-) Arabitol	3.04 g
K_2HPO_4	1.04 g
KH_2PO_4	0.54 g
Sodium taurocholate	0.29 g
$MgSO_4.7H_2O$	0.25 g
NH_4NO_3	0.16 g
Cycloheximide Solution	10.0 ml
Selenite Solution	10.0 ml
Crystal violet 0.1% solution	2.0 ml

Selenite solution

composition per 10.0 ml

NaOH	0.5 g
$Na_2SeO_3.5H_2O$	0.1 g

PREPARATION OF SELENITE SOLUTION : Add components to distilled/ deionized water and bring volume to 0.01 mix thoroughly. Filter sterlize.

Cycloheximide Solution –

Composition per	10.0 ml
Cycloheximide	0.02 g

PREPARATION OF CYCLOHEXIMIDE SOLUTION : Add cycloheximide to distilled/deionized water and bring volume to 10.0 ml. Mix thoroughly. Filter sterilize.

PREPARATION OF MEDIUM : Add components, except cycloheximide solution and selenite solution to distilled / deionized water and bring volume to 1.0 l. Mix thoroughly. Distribute 100.00 ml flasks. Gently heat and bring to boiling. Autoclave for 15 min at 15 psi pressure – 121^0C. Cool to 50^0C.

Aseptically add per 100 ml of medium, 1.0 ml sterite selenite solution and 1.0 ml cycloheximide solution. Mix thoroughly. Aspetically pour into sterite petri dish.

USE : For selective cultivation of *Agrobacterium tumefaciens* biovar.

Agrobacterium tumefaciens **selective medium**

Agar	15.0 g
Erythritol	3.05 g
K_2HPO_4	1.04 g
KH_2PO_4	0.54 g
Sodium taurocholate	0.29 g
$MgSO_4.7H_2O$	0.25 g
Cycloeximide solution	10.0 ml
Selenite solution	10.0 ml
Malachite green (0.1% soln.)	5.0 ml
Yeast extract (1% soln.)	1.0 ml

Selenite Solution

Preparation of selenite soultion

Cycloheximide solution

Preparation of Cycloheximide solution

Preparation of medium

USES : For selective cultivation of *Agrobacterium tumefaciens* biovar 2.

Agarobacterium Medium

Composition per liter

Agar	20.0 g
Mannitol	10.0 g

$NaNO_3$	4.0 g
$MgCl_2$	2.0 g
Calcium Propionate	1.2 g
$Mg_3(PO_4)_2$	0.2 g
$MgSO_4$	0.1 g
$MgCO_3$	0.075 g
$NaHCO_3$	0.075 g
Supplement	100.0 ml

pH 7.1 $\pm$ 0.2 at 25^0 C

(i) Collect the soil sample from rhizosphere soil of leguminous plants.

(ii) Serially dilute it ($10^2 - 10^7$) with sterile distilled water.

(iii) Pour aliquots of 1 ml of various dilution of soil to *Agrobacterium tumefaciens* selective medium and streak out the plates carefully or gently rotate the plate so as to spread the suspension on medium.

(iv) Incubate the plates at 37°C for 24 hrs. after incubation write down the colonial characteristics.

OBSERVATIONS :

Media used	Shape	Edge	Elevation	Opacity	Texture

(i) Collect the soil sample from rhizosphere of plant.

(ii) Make the dilution carefully.

M : Isolation of *Agrobacterium* sps. from crown gall.

    REQUIREMENTS : *Agrobacterium* agar medium plates, agrobacterium broth, test
e, nicron wire loop.

    PRINCIPLE : *Agrobacterium* is a rod shaped gram negative, soil borne, motile bacteria
en it grown on a variety of media, will exhibit difference in the macroscopic appearance
heir growth. These differences called cultured characteristics which are used as base for
arating Agrobacterium into taxonomic groups.

) Take infected part of plant tissues containing crown gall.

) Wash the plant tissues thoroughly first with tap water and then with sterile distilled
   water.

) Then, surface sterilize the plant part with sodium hypochlorite for 30 min. and then
   70% ethanol for one minute. Again wash with sterile distilled water 3-4 times to
   remove excessive chemicals.

) Transfer the plant part in sterite beaker containing small amount of Agrobacterium
   broth and then aseptically crush with glass rod or dissect by using nichrome blade.

) Prepare different dilutions.

) Pour 1 ml. suspension on Agropactesium agar medium and spread it by glass spreader.

OR

    Take a plate (Agrobacterium agar medium) and mark 4 sectors by glass marking
pencil on the back of plate.

) Sterilize the wire loop and take a loopful of culture under aspetic condition and streak
   out the plate by sector plate technique.

) Incubate the plates at 37^0C for 24 hrs. After incubation write down the colonical
   characteristics.

OBSERVATIONS :

Media used	Shape	Edge	Elevation	Opacity	Texture

PRECAUTIONS :

Prepare the Agrobacterium medium plates and broth carefully.
Experiment should be done under aseptic condition.

89

Differentiation of *Agrobacterium* cultural
characteristics from *Rhizobia*.

(*i*)   *Agrobacterium* can be differentiated from *Rhizobia* by a modified Nile blue test beca
only former organism reduce the dye.

(*ii*) On Yeast Extract Mannitol Agar (YEMA) medium members of *Rhizobaceae* g
well.

Some strains of *Agrobacterium* strongly absorb congo red or aniline blue from man
containing media. (YEM) YEMA with congo red containing medium colonie
Agrobacterium are pink, mucoid formation of star or rosette shaped aggregate
cells while Rhizobium utilizes congo red very slowly and form white colour, circ
and raised colonies.

(*iii*) On YEMA medium containing 2% salt conc. (NaCl) Agrobacterium is able to g
but Rhizobium cannot grow.

(*iv*) The changes in litmus milk are of diagnostic value. One group of strains (Biov
produces an alkaline reaction usually accompanied by the formation of a brown se
zone. Strains of biovar 2 produce an acid reaction (pink colour) sometimes accompa
by the formation of an acid clot.

(*v*)   On Glucose peptone agar medium Agrobacterium grows well but Rhizobium fai
grow on this

(*vi*) The vigorous and unusual oxidation of lactose to 3 ketolactose is the biochemical b
of the simple and specific diagnostic test for rapid differentiation of Agrobacter
strains. Agrobacterium when grow on lactose agar it utilises lactose by secreting
enzyme ketolactose but Rhizobia cannot utilise lactose. Colonies on such med
incubated for 4-10 days. Then Bendicts reagents are poured on plates. Formatio
yellow colour shows the presence of Agrobacterium.

Characteristics of *Agrobacterium*
on various media

	Media used	Indicator	Observations
1.	Yeast Extract Mannitol Agar (YEM)	Congo red	Abundent growth, short mucoid, pink colour colo The formation of star or ro shapped aggregates of c
2	Carbohydrate containing solid media (YEM) Yeast Extract Mannitol Agar		Circular, low convex to convex, mucous glisten opaque, white to beige col colonies with an entire e (dia. 2.4 mm).
3	Nutrient Agar		Moderate growth.
4	Agrobacterium tumefaciens selective media	Crystal Violent	Colonies are mucoid, circ low convex, opaque c agrobacterium will grow.

5	Agrobacterium tumefaciens selective media	Malachite green	Colonies are Circular mucoid low convex and only agrobacterium will grow.
6	Agrobacterium medium	–	Colonies are convex, circular, smooth, non pigmented to light beige.
7	Glycerophosphate agar	–	Production of white percepitate colonies are smooth, circular convex.
8	Triple sugar Iron agar (TSI)	Phenol red	NO H_2S production colonies mucoid, circular, smooth, convex.
9	Glucose pepton agar (GPA)		Only Agrobacterium will grow.

Differential Characteristics of *Agrobacterium*

	Characteristics	
1.	Growth temp. 25°C – 28°C	+
2.	Production of 3 ketolactose.	+
3.	Formation of acid from - meso-Erythritol	–
	Melezitose	+
4.	Formation of alkali from Na molonate	–
5.	Simmons citrate supplemental with 0.005% yeast extract (In biovar I- and in a biovar 2 +)	–/+
6	Reaction in litmus milk, Alkaline acid.	+/–
7.	Formation of pellicle in ferrous ammonium citrate solution.	+
8.	Growth factor requirements biotin and L-glutamic acid.	+ Biovar 2

Biochemical test for identification of Agrobacterium

	Characteristics	Test Result
1.	Oxidase test (disc)	+
2.	Indole formation	–
3.	H_2S formation on Triple sugar iron agar	–
4.	Acid production from D-Glucose, D-Fructose, L-arabionose, D-Xylose, mannitol, lactose, maltose, sucrose	+
5.	Gas from glucose	–
6.	Production of white precipitate on glycerophyosphate.	+
7.	Catalase test (By adding agar) H_2O_2 on colony	+
8.	Growth on sucrose + salt medium	+
9.	Hydrolysis of Gelatin, Casein, starch	–

Experiment 32

AIM : Demonstration of β-glucuronidas (gus) activity by histochemical assay.

REQUIREMENTS : Formaldehyde (0.3% v/v) X – GlCA Solution.

10 mM EDTA (pH 8-0)

100 mM Sodium phosphate

0.5 mM potassium ferrocyanide

0.1% (v/v) Trition X-100

0.5 mg/ml X-GlgA (5 bromo – 4 – chloro – 3 – indolyl B-D glucuronide)

Dissolve a appropriate amount of X-Glc-A in small volume of DMSo before addin to buffer.

Mannitol (0.3 M)

MES (10 mM, pH 5.6)

$NaH_2 PO_4$ (50 mm)

Petriplates, beakers

PRINCIPLES : β-glucuronidase is most widely used reporter gene in transfec plant cells. Because, the endogenous level of this hydrolase are very low. The enzyme ha unique property that it cleaves many β-glucuronide linkages with high efficiency, includ those β-gucuronidase conjugated with fluorescent chromogenic substrate X-GlcA (5-bro 4 chloro-3 indolyl-β-D glucuronide) and histochemical tags that are easily visualized wit cells deposition of blue spots (precipitate) indicating the expression of reporter gene. number of blue spots per dish can be considered to be a best semi qualitative estimate transfection efficiency.

METHOD :

(*i*) Take callus transformed with reporter gus gene.

(*ii*) Remove the medium in which the plant cells or tissues are maintained and repl with just enough X-GlcA solution to cover the biological material.

In some cases the intensity of b-glucuronidase staining can be enhanced by fixing tissues before addition of the X-GlcA solution.

Then, incubate the section in 0.3% (v/v) formaldehyde 10 mM MES/0.3 M mann for 30-60 minutes.

Rinse the section several times with 50 mM NaHPO$_4$ and then cover it with X-GlcA
ning solution.

(*iii*) Incubate the plates for 24 hours at 37°C.

(*iv*) Observe under light microscope the bombarded cells. Count the number of blue
s or up the b-gluronidase expression plasmid are synthesizing the enzyme should be a
k blue colour.

RESULT : Deposition of blue precipitation / spots demonstrate the expression of gus
e activity.

PRECAUTIONS :

(*i*) Care should be taken for preparing media and solutions for experiment.

(*ii*) β-glucuronidase is very sensitive to cation and to agar, which should be removed
n the stating vessel.

(*iii*) β-glucuronidase is senstitive to glutaraldehyde fixation and overfixing can lead
iisleading staining patterns.

AIM : **Demonstration of DNA polymorphism Restriction Fragment Len**
Polymorphism (RFLP) of different crop varieties.

REQUIREMENTS : Different crop varieties, restriction enzyme, Assay buf
Agarose, 50 XTAE, gel loading dye, vails.

PRINCIPLE : Here, DNA fragment of different length produced by cutting v
restriction enzymes. The technique employed for obtaining molecular finger print of
particular strain or species. A plant or organisms DNA is the blue print of its characteris
and is useful for obtaining genetic diversity, detection of allelic varient, polygenic traits
also used for comparative genome mapping. Detection was done by southern blot
technique.

METHOD :

(*i*) Take different crop varieties.

(*ii*) Isolate their DNA / or DNA samples provided.

(*iii*) Digest their DNA with restriction enzyme.

(*iv*) Run the samples on 1% agarose gel for 1-2 hrs. at 100 volts.

(*v*) Locate the bands of DNA fragments.

(*vi*) Transfer the DNA fragments of agarose gel onto nitrocellulose membrane
southern blotting). This is done by placing the gel on top of a buffer saturatted filter pa
then laying the nitrocellulose filter paper on top of the gel and finally placing some
filter paper on top of this membrane.

(*vii*) The DNA becomes tapped in the nitrocellulose membrane. Single stranded D
has a high affinity for nitrocellulose filter membrane.

(*viii*) Now the membrane is placed in a solution of radioactive 55 DNA o
oligodeoxynucleolide called probe.

(*ix*) Then, observe the fragments of DNA hybridized by probe with the hel
autoradiography.

(*x*) The membrane is now placed in close contact with an X-ray film and incubate
a desired period to allow images due to radioactive probes which is formed on the X

ı. The film is then develop to reveal distinct band(s) indicating position in gel DNA fragments
: are complementary to the radioactive probe used in the practical.

OBSERVATIONS : Compare RFLP maps of samples.

PRECAUTIONS :

(*i*) Enzyme·are temp. sensitive, hence place the vails containing enzyme on ice.

(*ii*) Ensure through mixing by gently tapping the vial, after addition of buffer and substrate
nzyme vial.

(*iii*) Use of fresh tip for each addition.

(*iv*) Care should be taken for preparing 1% agarose gel and 1XTAE buffer.

In RELP method there is a transfer of DNA fragments that have been cleaved by
ng restriction enzymes and which is separated by electrophoresis, onto nitrocellulose or
on paper of subsequently hybridized with a specific radiolabeled DNA probe.

FLOW CHART FOR RFLP

Crop Varieties

↓

Isolation of DNA

↓

Digest the DNA with restriction enzyme

↓

Gel electrophoresis

↓

Southern blotting

↓

Autoradiography

↓

DNA bands seen which shows RFLP pattern

AIM : Demonstration of DNA polymorphism through Random Amplified Polymorp DNA (RAPD).

REQUIREMENTS : Crop varieties, Taq DNA polymerase, deoxynucleot triphosphate, Assay buffer, 1% gelatin DNA template, primer (forward and reverse), agaro nuclease free water, gel loading dye, 50XTAE, PCR tube.

PRINCIPLE : RAPD is a PCR based technique for detection of polymorphism DNA level. Here, a single short (10 mers) synthetic nucleotide primers are used in PC PCR selectively amplifies the specific segments of DNA of an orgs/plants. An orgs or pl DNA contains the blue print of its characteristics. In case of plants this would inclu features like yield, drought, resistance, starch content and so on. It gives information c strain or species.

METHOD :

(*i*) Take different crop varieties.

(*ii*) Isolate their DNA / or different samples of DNA provided.

(*iii*) DNA amplification has been done by using PCR technique. (Which was discuss in PCR experiment no.)

(*iv*) Run the sample on 1% agarose gel for 1-2 hrs. at 100 volts.

(*v*) Locate the amplified product by comparing with fragments of different sample

OBSERVATIONS : Compare RAPD map of samples.

FLOW CHART FOR RAPD

Crop Varieties

↓

Isolation of DNA

↓

PCR

↓

Gel Electrophoresis

↓

Amplified product seen which shows RAPD pattern

PRECAUTIONS :

(*i*) Carefully add the required amount of reagent into a PCR tube.

(*ii*) Carry out the amplification using proper reaction condition (time and temperature)

(*iii*) Preincubate at 95°C for 5 min in the absence of enzyme to inactivate harmful nuclease or protease in the sample.

VIVA VOCE

Q.1. What is size of *E.coli*.

Ans. *E.coli* is a Gram negative bacteria, size –1.3 × 0.40 microns arranged singly or in pairs and clusters.

Q.2. What are the Cultural characteristics of *E.coli*.

Ans. *E.coli* grown on nutrient agar with large, thick, grayish white, smooth, transluscent colonies.

On Mac Conkey agar the colonies are bright pink because of lactose fermentation.

On E.M.B. agar they produce colonies with dark centre and greenish metallic sheen.

Q.3. Is *E.coli* are pathogenic organism.

Ans. Yes, it is an opportunistic pathogen. under certain condition it causes.

1. Diarrhoea or gastroentritis.
2. Urinary tract infection.
3. Pyogenic infections viz peritonitis cholecystis and meningitis.

Q.4. What are the special characteristics of E.M.B. and Mac Conkey agar. medium.

Ans. E.M.B. is selective and differential medium that allows the growth of the members of Enterobacteriaceace and also identifies the lactose fermenters. Other orgs. are inhibited by the bye. The medium is specially used to differentiate between *E.coli* and Klebsicella. (Enterobacter).

E.coli – which is a strong fermenting organism produces large amount of acid. Resulting in precipitation of eosion and methylene blue to which colonies of *E.coli* have greenish metallic sheen. with dark centre.

Mac Conkey agar- Bite salt content of the medium inhibits the growth of gram positive bacteria, thus the medium is selective for growth of gram negative bacteria.

Lactose and pH indicator (Neutral red) makes it differential. Lactose fermenting bacteria produce acid and due to acid formation indicator gives pink colour, Lactose non fermenting bacteria gives colourless (Pale) colonies.

Q.5. What are the uses of nutrient agar and nutrient broth.

Ans. Both are used for general cultivation of bacteria. They are designed to support the growth of wide range of bacteria.

Q.6. What are the properties of various components of N agar.

Ans. 1. Peptone- Nitrogen source and buffer.

2. Meat extract- Supplies growth factors (Mineral salts and amino acids.)

3. Sod. chloride- Maintains osmotic pressure.

4. Agar-agar-Solidifying agent.

Q.7. Define turbidity and how will you measured the turbidity.

Ans. It is the method of estimation of bacterial cell growth or population in broth, culture or aqueous suspension by the measurement degree aqueous of opacity or turbidity of the suspension, here, the growth is determine by photometric method in terms of optical density. Where the basic principle is that the optical density is directly proportional to the micro orgs. present.

Q.8. Which type of growth curve is found in bacteria.

Ans. Sigmoidal growth curve.

Q.9. If no colorimeter is available can you observe turbidity.

Ans. With naked eyes and prepare the following scale as the index of turbidity.

1. Maximum turbidity (Max growth) +++

2. More turbidity (more growth) ++

3. Low turbidity (least growth) +

4. No turbidity. (No growth) –

Q.10. Is there any other instrument which measure turbidity.

Ans. Turbidometer.

Q.11. Define growth of growth curve.

Ans. Growth means the increase in population of micro orgs.

Growth curve is a graphic representation of the changes of the population of bacteria in a culture medium.

Q.12. Can you trace the growth.

Ans. Yes, growth can be traced by-

1. Cell count-   Directly- Microscopy, Eletronic particle counter.
                 Indirectly- Colony count.

2. Cell mass-   Directly- Measurement of cell N_2, direct weighing
                 Indirectly- Turbidity.

3. Cell activity- Indirectly by relating the biochemical
                 activity to size of population.

Q.13. Define plasmids.

Ans. Plasmids are double stranded, circular, self replicating extra chromosomal DNA molecules. They are commonly used as cloning vectors in molecular biology.

Q.13. What are the steps to isolate plasmid D.N.A.

Ans. 1. Growth

2. Harvest and lysis of bacteria.

3. Isolation of plasmid D.N.A.

Q.14. What are the properties of various components of solution I, II, III & IV in plasmid isolation.

Ans. Solution-I

Glucose- prevents immediate lysis of cells. Tris (pH 8.0) maintain pH.

EDTA- Chelates metal icons & weakens cell wall.

Solution-II

NaOH - Denatures chromosomal and plasmid DNA.

SDS – Denatures bacterial proteins and disrupt – cell membrane.

Solution-III

Sodium acetate renatures plasmid D.N.A.

Solution-IV

Isopropanol-pptes plasmid D.N.A.

R Nase A- Degrades RNA without affecting D.N.A.

Q.15. State the name of plasmid.

Ans. You will purify pUC18 plasmid form D5 (*E.coli* strain).

Size – 2686 bp with ampicillin resistance gene.

Q.16. What are the basic steps of gel electrophoresis.

Ans. 1.   Preparation of agarose gel.

2.   Electrophoresis of the DNA fragment.

3.   Visualization of DNA.

Q.17. Define agarose.

Ans. Agarose is a linear polymer extracted from seaweeds.

Q.18. What is the charge present on DNA ?

Ans. DNA is negatively charged at natural pH.

Q.19. What is the basic principle of electrophoresis of DNA fragments.

Ans. Electrophoresis is a technique used to separate charged molecules. When electric filed is applied across the gel, DNA migrates towards the anode. Migration of DNA through the gel is dependent upon.

1.   Molecular size of DNA.

2.   Agarose concentration.

3.   Conformation of DNA.

4.   Applied current.

Q.20. How will you visualize the DNA fragments.

Ans. By staining with dye bromophenol blue. Alternatively Ethidium bromide can be used for visualizing DNA fragments. Add Ethidium bromide to motten agarose to final concentration of 0.5 mg/ml (from a stock of 10 mg/ml in water. When temp. is around 50°C mix and cast the gel. After electrophoresis of DNA samples can be visualized under uv light they appear fluorescent (Orange band).

.21. Name the two commonly used dyes for DNA visualization.

ns. Xylene cynol and Bromophenol blue. These dye migrate at the same speed as double stainded DNA of size 5000 bp and 3000 bp respectively.

.22. What is the charge present on tracking dye.

ns. Negative.

.23. How will you monitored the gel electrophoresis.

ns. The process of gel electrophoresis is monitored by observing the migration of visible dye (tracking dye) through gel.

.24. Define plating efficiency.

ns. It is a quantitative assessment of the percents of cell colonies per total number of cells or protoplasts plated at a defined density for a specific time period. It can be determined by counting cell colonies under microscope at the end of experiment and calculated by using the formula.

$$\text{Plating efficiency (PE)} = \frac{\text{No. of colonies/plate}}{\text{No. of cell Unit}} \times 100$$

.25. State the name of some vital stains used for viability testing of cells and protoplasts.

ns.

Stain used	Observation Method	Results
1 Evan's blue 0.5% in culture medium or in D/W.	Examine with bright filed microscopy.	Dead cells will stained blue living cells will be unstained.
2. Phenosafranin 0.1% culture medium	— " —	Dead cells will be stained in red, living cells will not be stained.
3. Fluorescein diacetate dissolve 5 mg in 1 ml of acetone dilute the soln. with culture medium to final concentration 0.01%	Examine with florescene microscopy using UV light.	Fluorescence will be seen only in living cells.

.26. Name some commonly used plant growth regulators.

ns. Auxin – Indole 3 acetic acid, (IAA) use for callus induction at 10-30 μM.

Indole 3 butyric acid (IBA) use for rooting shoots regenrated via organogenesis. (1.50 μM).

Cytokinin – 6 Furfurylam inopurine (Kinetin) K. culture media for callus induction, growth of callus and cell suspensions and introduction of morphogenesis (1-20 μM).

Cytokinin – 6 Bengylaminopurine BAP callus induction, growth of callus, rapid multiplication of shoots, buds.

Gibberellin GA_3 – Used in callus culture can promote shoot growth also used enhance development of embryovule cultures. (0.3-48 mM).

Abscisic acid ABA – Used at (0.04-10mM) to prevent precocious germination ai promote normal development of somatic embryos.

Q.27. Define explants.

Ans. The piece of tissue isolated from the intact plant that is used to initiate the culture

Q.28. What charge is present on the protoplast.

Ans. Plant protoplasts carry a negative surface charge.

Q.29. What are the properties of isolated protoplast.

Ans. Isolated protoplast is naked plant cell (without cell wall) surrounded by plasm membrane, potentially capable of cell wall regeneration, cell division, growth ai plant regeneration in culture.

Q.30. What are the methods of protoplast fusion.

Ans. 1.   Spontaneous fusion (Mechanical fusion)

2.   Induced fusion

(a)   Chemofusion (fusion induced by alcohol chemicals – P.E.G., Calcium io polyvinyl alcohol)

(b)   Electrofusion (fusion induced by electric current)

Q.31. Name the enzymes which are used in islation of protoplast.

Ans. Macerase (Macerozyme), Pectinase, Hemicellulase, Cellulysin, Rhozyme HP 15 Driselase.

Q.32. Define primary culture.

Ans. A culture started from explants taken directly from plant body.

Q.33. What is organogenesis.

Ans. Organogenesis means the development of adventitious organs or primordia fro undifferentiated cell mass in tissue culture by the process of differentiation.

Q.34. What is caulogenesis.

Ans. It is a type of organogenesis by which only adventions shoot bud initiation tak place in the callus tissue.

Q.35. What is Rhizogenesis.

Ans. Rhizogenesis is a type of organogenesis by which only adventitious root formatii takes place in the callus tissue.

Q.36. What is meristemoids.

Ans. Meristemoid is a localised group of meristematic cells that arise in the callus cultu and may give rise to shoots and/or roots.

Q.37. What is somatic embryogenesis.

Ans. In plant tissue culture, the development pathway of numerous well organised, sm: embryogenic potential somatic plant cell of callus tissue or cells of suspension cultu is known as somatic embryogenesis.

38. What is embryogenic potential.

s. The capability of somatic plant cell of culture to produce embryoids is known as embryogenic potential.

39. What do you understand by organ culture.

s. The maintenance or growth of organ primordia in vitro without forming callus tissue.

40. What is embryoid.

s. Embryoid is a small, well organised structure comparable to the sexual embryo, which is produced in tissue culture of dividing embryogenic potential somatic cells.

41. What is the difference between direct and indirect embryogenesis.

s. Direct embryogenesis i.e. cells of explant undergo direct embryogenesis from proembryonic determined cells in absence of callus proliferation.

Indirect embryogenesis – i.e. cells of explant first undergo callus proliferation and embryoids develop within the callus tissue from induced embryogenic cells.

42. Name the factors affecting somatic embryogenesis.

s. Chemical factor – Auxin, Cytokinin.

43. What is Cytodifferentiation.

s. Cytodifferentiation means the invitro vascular differentiation particularly the xylem elements within the callus tissues.

44. What is androgenesis.

s. Androgenesis is the in vitro development of haploid plants originating from totipotent pollen grains through series of cell division and differentiation.

45. Define Haploid.

s. A cell or nucleus containing a single set of chromosomes.

46. What are the modes of androgenesis ?

s. Two modes of androgenesis.
1. Direct androgenesis
2. Indirect androgenesis
1. **Direct androgenesis** – In this type, microspore behave like a zygote and undergoes change to form embryoid whichultimately give rise to a plantlet.
2. **Indirect androgenesis** – The microspore, instead of ndergoing embryogenesis divide, repeatedly to form a callus tissue which differentiates into haploid plantlets.

47. Is cell division required for xylem differentiation.

s. Two views
1. Tracheary element differentiation many be related to cell cycling activety, although some cells differentiatedirectly without any expression of mitotic division.
2. Cell division is not required for xylem differentiation, Inibitor studies with caffeine and colchicine have provided evidence that would xylem formation in Pisum roots is a direct differentiation process.

Q.48. What are the factors affecting cytodifferentiation.

Ans. Chemical factors – Auxin, Cytokinin, Gibberellin,

Physical factors –　1.　Light (inhibitory effect in xylogenesis)

2. Temp., pressure and water stress.

Q.49. What is totipotency.

Ans. Totipotency is the genetic potential of a plant cell to produce the entire plant. In oth
words, toptipotency is the cell characteristic in which the potential for forming a
the cell types in the adult organism is retained.

Q.50. Are all cells in culture are totipotent.

Ans. Limited expression of totipotency and it vary from plant to plant.

- Variation of chromosome number in the cells of callus tissue is one of the ma
factors that causes the limited expression of totipotency. This variability may be d
to either pre-existing variation in the somatic cells of the explant (genetic) or variati
generated during tissue culture (epigenetic). Changes in chromosome number a
aneuploid, polyploid etc. As a result, a mixoploid callus tissue is formed in t
subsequent growth. But very often, from these mixoploid callus culture
organogenesis and/or embryogenesis occur mostly from diploid cells. Therefore, a
cells of the callus tissue are not able to express their totipotency.

Q.51. What is the importance of totipotency in plant science.

Ans. -　Reconstruction of plants from the totipotent cells.

-　The totipotentiality of somatic cells has been exploited in vegetative propagati
of many economical, medicinal as well as agriculturally important plant specie

-　Genetic modification of plants, production of homozygous diploids plant throu
haploid cell culture, somatic hybridization etc.

Q.52. How the callus tissue is formed.

Ans. Formation of callus tissue is the outcome of cell expansion and cell division of t
cells of the explant.

- Exogenously supplied hormones.

Q.53. What are the characteristic of callus.

Ans. 1.　Internal & External morphology, shape and size.

2.　Texture – on the basis of texture, callus tissue is divided into two categories.

(*i*)　Soft callus

(*ii*)　Hard callus

(*i*)　Soft callus is friable in nature and is made of heterogeneous mass of ce
having minimal contact.

(*ii*)　Hard callus consists of giant cells, tracheid like like cells and closely pack
cells i.e. compact in nature

– Hard callus may be nodular in from.

3. Colouration – Generally, callus tissue is sometimes may be pigmented. Pigmentation may be uniform or patchy. Callus tissue many be green in colour.

   – Green colour develops due to development of chloroplastid in the cells of callus tissue e.g. callus tissue from the cotylendons of soyabean.

   – Yellow colour possibly due to synthesis of carotenoid pigments e.g. Nigella sativa.

   – In some cauliflower cutlure, callus tissue is purple in colour due to accumulation of anthocyanin in vacuoles or due to production of oxidized form of 3, 4 dihydroxy phenylalanine.

4. Brown colour – frequently develops in the explant and subsequently in the callus tissue. This is mainly due to excretion of phenolic substances.

Q.54. What are the importance of callus culture.

Ans. - Somatic embryo development.

   - Regeneration of plant from genetically variable cells of the callus tissue.

   - Useful to obtain commercially important secondary metabolites.

   - Biochemical assays can be performed from callus culture.

Q.55. Define homokaryon and hereokaryon.

Ans. Homokaryon – A cell will two or more identical nuclei as a result of fusion & a cell containing two or more nuclei unlike genetic make up are called heterokaryon.

Q.56. State the first and most commonly employed thermostable DNA polymerase.

Ans. Taq DNA polymerase which was isolated from bacterium Thermus aquaticus found in the hot springs of yellow stone National Park.

Q.57. Name some thermostable DNA polymerases.

Ans. Vent^{Tm} isolated from Thermococcus litoralis. UITma^{Tm} isolated from Thermotoga maritima. Tth isolated from Thermus thermophilus P7U DNA polymerase isolated from a marine bacterium.

Q.58. Who is the inventor of PCR technique.

Ans. Kary Millis

Q.59. What are the basic aspects of the PCR.

Ans. PCR is a simple method for in vitro enzymatic amplification of specific nucleic acids using multiple cycles of template denaturation, primer annealing and primer elongation.

Q.60. What are the basic components of the PCR.

Ans. Template DNA, primers (forward primer and reverse primer) Taq DNA polymerase.

Q.61. What are the other components of the PCR.

Ans. $MgCl_2$ concentration. It is a critical component because it is not only required by the DNA polymerase for efficient activity but also forms a soluable complex with the dNTPs, which is essential for incorporation in the extension step of the PCR cycle.

Q.62. Define DNA amplification.

Ans. Multiplication of a piece of a DNA in a test tube into many thousand or million of copies by using PCR.

Q.63. What are the applications of PCR.

Ans. RT – PCR is used for analysis of differential gene expression and in some cases as a tool for diagnosis of several retro viruses e.g. Hepatitis C.

- Molecular cloning of PCR products.
- Multiplex PCR is used in the diagnosis of several genetic disorders such as Duechine muscular dystrophy. (DMD)
- Used as a tools for detecting the polymorphisms.
- Used in vitro mutagenesis.
- In medical diagnosis for detection and quantify minute amount of a pathogen.

Q.64. State the function of restriction endonuclease.

Ans. Restriction endonucleases are the bacterial enzymes that cut double standed DNA molecules in a precise and reproducible manner. Restriction enzymes recognize specific sequences in double stranded DNA and cleave the DNA, usually within the recognition site, to yield fragments of defined length.

Q.65. Name some restriction endonucleases enzyme and the micro organisms from which they are isolated.

Ans.	Enzyme	Organism
	Eco RI	Escherichia coli
	Bam HI	Bacillus amyloliquefaciens
	Bg / II	Bacillus globigii
	Hind III	Haemophilus influezae Rd
	Hae III	Haemophilus aegypaticus
	Kpn I	Klebsiella pneumonie
	Sma I	Serratia marcescens

Q.66. Define RAPD.

Ans. Random amplified polymorphic DNA. RAPD is a PCR based technique for detecting polymorphism at DNA level.

Q.67. Define RFLP.

Ans. Restriction fragment length polymorphism. In this technique DNA fragment of different length produced by cutting with restriction enzymes.

Q.68. Define restriction map.

Ans. A linear array of sites on DNA cleaved by various restriction enzymes.

Q.69. What do you understand by Southern blotting.

Ans. A method for transferring denatured DNA molecules which is separated by gel electrophoresis onto a nitrocellulose membrane on which hybridization can be done.

Q.70. What are artificial seeds.

Ans. Artificial seeds are the living seeds like structure which are made experimentally by a technique where somatic embryoids derived from plant tissue culture are encapsulated by a hydrogel and such encapsulated embryoids behave like true seeds if grown in soil and can be used as a substitute of natural seeds.

Q.71. Define micropropagation.

Ans. The asexual or vegetative propagation of whole plant using tissue culture technique is known as micropropagation

OR

Regeneration of whole plant through tissue culture is called micropropagation.

Q.72. What do you understand by somaclonal variation.

Ans. It is the genetic variability which is generated during tissue culture.

Q.73. State the name of hydrogel used in making artificial seeds.

Ans. Sodium alginate, Gel-rite^{TM}, Sodium alginate with gelatin, Carragenan with locust beam gum.

Q.74. Define reporter gene and give some examples of it.

Ans. A gene encoding a product that can readily be assayed during genetic transformation. It may be connected to any promoter of interest so that it may express.

e.g. GUS, Chloramphenicol transacetylase, lac Z, lux, nos, OCS gene.

Q.75. Name the enzyme expressed by GUS gene.

Ans. β-glucuronidase.

Q.76. Give The examples of some of the reporter genes used as screenable markers.

Ans.

Reporter gene	Enzymes expressed
cat	Chloramphenicol acetyl transferase
lac Z	β-galactosidase
lux	luciferase
nos	Nopaline synthase
npt II	Neomycin phosphotransferase
ocs	Octopine synthase

Q.77. What is the source of GUS gene and how it will be detected.

Ans. *E.coli* is the source of GUS gene and it will be detected by a range of histochemical (such as X-Gluc), chromogenic or fluorescent substrates and can be used to assay activity.

References

1. Atlas Ronald M. (2004)

   Handbook of Microbiological Media (Third Edition) CRC Press London, New York, Washington DC

2. Celis Julio E.

   Cell Biology – A laboratory Handbook volume one second edition.

   Academic Press, London, New York, Tokyo.

3. Dixon R.A. and R.A. Gonzales (2004)

   Plant cell culture – A practical approach second edition.

   Oxford University Press Oxford New York, Tokyo

4. De Kalyan Kumar

   An Introduction to Plant Tissue culture (1995)

   New Central Book Agency Calcutta.

5. Endress Rudolf (2004)

   Plant cell Biotechnology.

   Springer (India) Private Limited New Delhi

6. Kried Noel R and Holt John G. (1984)

   Bergey's Manual of Systematic Bacteriology vol. 1

   Williams and Wilkins

   United States of America.

7. Narayanaswamy S. (2002)

   Plant cell and Tissue culture.

   Tata McGraw Hill Publishing Company Limited New Delhi (India)

8. Rapley Ralph and John M. Walker (1998)

   Molecular Biomethods Handbook.

   Humana Press, Totowa, New Jersey.

9. Smith Roberta H. (2005)
    Plant Tissue Culture
    Techniques and Experiments second edition
    Academic Press An Imprint of Elsevier New Delhi India.

10. Slater Adrian, Nigel Scott and Mark Fowler (2003)
    Plant Biotechnology
    The genetic manipulation of plants.
    Oxford University Press Oxford New York.

11. Sambrook Jospeh and Russell David W. (2001)
    Molecular cloning – A Laboratory Manual Volume 3
    Third edition, International al edition Cold Spring Harbor Laboratory Press New York.

INDEX

A

Abscisic acid, 53
Agrobacterium, 86, 89, 91
Androgenesis, 70, 103
Artificial seeds, 79
Autoclave, 1
Auxin, 52

B

Bacillus Amyloliquefaciens, 106
Bacillus globigii, 106
Bacteriophage titration, 39
Bromophenol blue, 101

C

Callus Culture, 62, 66
Caulogenesis, 102
Cell suspension culture, 68
Centrifuge, 10
Cell protoplast washing media, 55
Chlorocholine Chloride, 53
Colorimeter, 9
Cytodifferentiation, 103
Cytokinin, 52

D

DNA amplification PCR, 49, 105
DNA extracting buffer, 19
DNA isolation plant, 84
DNA (genomic DNA), 33
isolation (bacteria)
DNA molecular size, 35

E

E.coli, 25, 26, 98
EDTA, 20, 101

E (continued)

Embryogenic potential, 102
Eosin Methylene Blue
agar (EMB), 18
Ethylene, 53
Evan's blue, 101
Explant, 56, 102

F

Fluorescein diacetate, 58, 101

G

Gel electrophoresis. 14
Gibberellin. 52
Glucuronidase (GUS), 92
Growth Curve, 99
Growth Characteristics, 25
Growth index, 78

H

Haploid, 103
HEPA, 5
Hemicellulase, 102
Heterokaryon, 104
Hot air oven, 3
Homokaryon, 104

I

Incubator, 5
IPTG, 45

K

Kary Millis, 105
Klebsiella pneumonie, 106

Notes

Notes

Notes

Notes